Synthesis Lectures on Digital Circuits & Systems

Series Editor

Mitchell A. Thornton, Southern Methodist University, Dallas, USA

This series includes titles of interest to students, professionals, and researchers in the area of design and analysis of digital circuits and systems. Each Lecture is self-contained and focuses on the background information required to understand the subject matter and practical case studies that illustrate applications. The format of a Lecture is structured such that each will be devoted to a specific topic in digital circuits and systems rather than a larger overview of several topics such as that found in a comprehensive handbook. The Lectures cover both well-established areas as well as newly developed or emerging material in digital circuits and systems design and analysis.

Steven F. Barrett

Arduino V: Machine Learning

 Springer

Steven F. Barrett
Department of Electrical and Computer
Engineering
University of Wyoming
Laramie, WY, USA

ISSN 1932-3166   ISSN 1932-3174 (electronic)
Synthesis Lectures on Digital Circuits & Systems
ISBN 978-3-031-21879-8   ISBN 978-3-031-21877-4 (eBook)
https://doi.org/10.1007/978-3-031-21877-4

This Springer imprint is published by the registered company Springer Nature Switzerland AG
The registered company address is: Gewerbestrasse 11, 6330 Cham, Switzerland

Preface

This book is about the Arduino microcontroller and the Arduino concept. The visionary Arduino team of Massimo Banzi, David Cuartielles, Tom Igoe, Gianluca Martino, and David Mellis launched a new innovation in microcontroller hardware in 2005, the concept of open-source hardware. Their approach was to openly share details of microcontroller-based hardware design platforms to stimulate the sharing of ideas and promote innovation. This concept has been popular in the software world for many years. In June 2019, Joel Claypool and I met to plan the fourth edition of "Arduino Microcontroller Processing for Everyone!" Our goal has been to provide an accessible book on the rapidly evolving world of Arduino for a wide variety of audiences including students of the fine arts, middle and senior high school students, engineering design students, and practicing scientists and engineers. To make the book even more accessible to better serve our readers, we decided to change our approach and provide a series of smaller volumes. Each volume is written to a specific topic and audience. This book, "Arduino V: AI and Machine Learning", explores Arduino applications in the fascinating and rapidly evolving world of small, local microcontroller-based AI and ML applications. The first three chapters explore the Arduino IDE, the Arduino Nano 33 BLE Sense, and sensor and peripheral interface techniques. In the remaining three chapters, we take a tutorial approach to Artificial Intelligence (AI) and Machine Learning (ML) concepts appropriate for implementation on a microcontroller including: K Nearest Neighbors (KNN), Decision Trees, Fuzzy Logic, Perceptrons, and Artificial Neural Nets (ANN).

Approach of the Book

This book is part of a multi-volume introduction to the Arduino line of processors. The book series also serves as the "fourth edition" of "Arduino Microcontroller Processing for Everyone!" When discussing plans for a fourth edition, Joel Claypool and I (sfb) decided to break the large volume up into a series of smaller volumes to better serve the needs and interests of our readers. I have tried to strike a balance between each volume being independent of one another while holding to a minimum of information

v

contained in other volumes. For completeness and independence, this volume contains tutorial information on getting started, microcontroller interface information, and motor control partially contained in some of the other volumes and related works completed for Morgan and Claypool and Springer Nature. I have identified via chapter footnotes the source of this information contained elsewhere in the series. The book series thus far includes:

- "Arduino I: Getting Started"
- "Arduino II: Systems"
- "Arduino III: Internet of Things"
- "Arduino IV: DIY Robots—3D Printing, Instrumentation, Control"
- "Arduino V: AI and Machine Learning"

In this book, "Arduino V: AI and Machine Learning", we concentrate on Artificial Intelligence (AI) and Machine Learning (ML) applications for microcontroller-based systems. In a recent release, the Arduino Team stated "Arduino is on a mission to make machine learning simple enough for everyone to use [1]". Those acquainted with AI and ML concepts might counter these concepts are most appropriate for more powerful computing platforms. However, recent developments have allowed certain AI and ML applications to be executed on microcontrollers once they have been trained. There are applications that lend themselves to remote, battery-operated microcontroller-based AI and ML applications [3]. In this book, we limit our discussions to AI and ML techniques specifically for microcontrollers. The intent is to introduce the concepts and allow you to practice on low cost, accessible Arduino hardware and software. Hopefully, you will find this book a starting point, an introductory, to this fascinating field. We provide a number of references for further exploration.

Figure 1 illustrates the relationship between Artificial Intelligence, Machine Learning, and Deep Learning. The goal of Artificial Intelligence is for computing machinery to imitate and mimic intelligent human behavior. Some trace the origins of AI back to 1300 BCE [6]. We limit our historical review to AI developments within the 20th century and forward. Within the realm of AI, we explore Fuzzy Logic.

Machine Learning is under the umbrella of Artificial Intelligence. Its goal is to develop algorithms to control a process or categorize objects. The developed algorithm undergoes a learning step where input data is used to confirm or develop desired controller outputs. During the learning process the algorithm adjusts certain weights to improve the performance of the algorithm. Within the realm of ML we explore K Nearest Neighbor (KNN) algorithms, decision trees, perceptrons, and Artificial Neural Networks (ANN). Deep Learning involves the development of algorithms using multi-layer Artificial Neural Networks (ANN).

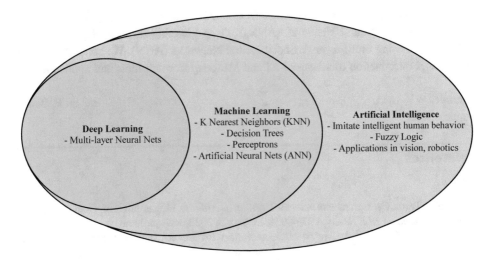

Fig. 1 Artificial Intelligence and Machine Learning [5].

As in other books in the series, for completeness, we provide prerequisite information in Chap. 1. The chapter provides a Quickstart guide to getting started with the Arduino Integrated Development Environment (IDE).

Chapter 2 introduces the Arduino Nano 33 BLE Sense microcontroller development board. **This is a 3.3 VDC microcontroller.**[1] The Nano hosts the NINA B306 module which includes a powerful 64 MHz, 32-bit Arm Cortex-M4F Nordic Semiconductor nRF52840 processor with 256 KB of static RAM (SRAM) and 1 MB of flash memory. The module also contains Bluetooth and Zigbee communications, serial communication subsystems (UART, I2C, SPI), direct memory access features, analog-to-digital converters (ADC), and a 128-bit Advanced Encryption Standard (AES) co-processor [2, 4].

Onboard the Nano 33 development board is an extensive series of peripherals including a nine axis inertial measurement unit; sensors for barometric pressure, temperature, humidity, proximity, light, and gesture detection; a digital microphone; and a cryptographic co-processor [2].

Chapter 3 introduces the extremely important concept of the operating envelope for a microcontroller. The voltage and current electrical parameters for the Arduino microcontrollers are presented and applied to properly interface input and output devices to the **3.3 VDC** Arduino Nano 33 BLE Sense microcontroller development board. We provide the bare essentials of interfacing to the Nano 33 for the applications discussed in the book.

In Chap. 4, following a brief historical review, we explore the Machine Learning concepts of K Nearest Neighbors (KNN) and Decision Tree classification techniques. Within

---

[1] We emphasize throughout the book that this is a 3.3 VDC processor. Processor inputs and outputs must not exceed 3.3 VDC!

the realm of AI, we explore Fuzzy Logic in Chap. 5. In Chap. 6, we explore the perceptron and Artificial Neural Networks (ANNs). Deep Learning involves the development of algorithms using multi-layer Artificial Neural Networks (ANN). We conclude Chap. 6 with a brief introduction to advanced AI and ML deep learning tools and applications.

Laramie, WY, USA   Steven F. Barrett
January 2023

References

1. Arduino Team, *Get started with machine learning on Arduino*, blog.arduino.cc, October 15, 2019.
2. *Arduino Nano 33 BLE Sense,* ABX00031, January 5, 2022, www.arduino.cc
3. G. Lawton, *Machine Learning on Microcontrollers Enables AI,* targettech.com, November 17, 2021.
4. *nRF52840 Advanced multi--protocol System--on--Chip*, nRF52840 Product Brief Version 1.0, Nordic Semiconductor.
5. J.P. Mueller and L. Massaron, *Artificial Intelligence for Dummies,* John Wiley and Sons, Inc, 2018.
6. C. Pickover, *Artificial Intelligence an Illustrated History*, Sterling, New York, 2019.

Acknowledgments

A number of people have made this book possible. I would like to thank Massimo Banzi of the Arduino design team for his support and encouragement in writing the first edition of this book: "Arduino Microcontroller: Processing for Everyone!"

I would also like to acknowledge Joel Claypool for his publishing expertise and support to a number of writing projects. His vision and expertise in the publishing world has made this book possible. Joel "retired" in September 2022 after 40 plus years of service to the U.S. Navy and the publishing world. On behalf of the multitude of writers you have provided a chance to become published authors, we thank you! The next adventurous chapter in Joel's life begins with an upcoming hurricane relief effort service trip. I dedicate this book to you my friend.

I would also like to thank Dharaneeswaran Sundaramurthy of Total Service Books Production for his expertise in converting the final draft into a finished product.

Finally, as most importantly, I would like to thank my wife and best friend of many (almost 50) years, Cindy.

Laramie, WY, USA   Steven F. Barrett
January 2023

Contents

1 Getting Started 1
  1.1   Overview 1
  1.2   The Big Picture 1
  1.3   Arduino Quickstart 3
       1.3.1   Quick Start Guide 3
       1.3.2   Arduino Development Environment Overview 6
       1.3.3   Sketchbook Concept 7
       1.3.4   Arduino Software, Libraries, and Language References 7
       1.3.5   Writing an Arduino Sketch 7
  1.4   Application: LED Strip 11
  1.5   Summary 14
  1.6   Problems 15
  References 15

2 Arduino Nano 33 BLE Sense 17
  2.1   Overview 17
  2.2   Arduino Nano 33 BLE Sense SoC Board 18
  2.3   Arduino Nano 33 BLE Sense Features 19
  2.4   NINA B306 Module Subsystems 20
       2.4.1   B306 Module Memory 21
  2.5   NINA B306 Module Peripherals 22
       2.5.1   Pulse Width Modulation (PWM) Channels 22
       2.5.2   Serial Communications 23
       2.5.3   Bluetooth Low Energy (BLE) 35
  2.6   Nano 33 BLE Sense Peripherals 43
       2.6.1   Nine Axis IMU (LSM9DS1) 44
       2.6.2   Barometer and Temperature Sensor (LPS22HB) 46
       2.6.3   Relative Humidity and Temperature Sensor (HTS221) 47
       2.6.4   Digital Proximity, Ambient Light, RGB, and Gesture
              Sensor (APDS–9960) 49

      2.6.5   Digital Microphone (MP34DT05) 55
2.7   Application: Bluetooth BLE Greenhouse Monitor 57
2.8   Summary 64
2.9   Problems 64
References 65

3 Arduino Nano 33 BLE Sense Power and Interfacing 67
3.1   Overview 67
3.2   Arduino Power Requirements 68
3.3   Voltage Regulators 68
      3.3.1   Powering the Nano 33 From Batteries 68
3.4   Interfacing Concepts 68
3.5   Input Devices 70
      3.5.1   Switches 70
3.6   Output Devices 72
      3.6.1   Light Emitting Diodes (LEDs) 72
      3.6.2   Serial Liquid Crystal Display (LCD) 73
3.7   Motor Control Concepts 73
      3.7.1   DC Motor 75
3.8   Application: Dagu Magician Robot 78
      3.8.1   Requirements 82
      3.8.2   Circuit Diagram 82
      3.8.3   Dagu Magician Robot Control Algorithm 83
      3.8.4   Testing the Control Algorithm 92
3.9   Summary 92
3.10   Problems 93
References 93

4 Artificial Intelligence and Machine Learning 95
4.1   Overview 95
4.2   A Brief History of AI and ML Developments 97
4.3   K Nearest Neighbors 99
4.4   Decision Trees 104
4.5   Application: KNN Classifier 118
4.6   Application: Decision Trees 118
4.7   Summary 121
4.8   Problems 121
References 121

5 Fuzzy Logic 123
5.1   Overview Concepts 123
5.2   Theory 124
      5.2.1   Establish Fuzzy Control System Goal, Inputs, and Outputs 126

        5.2.2   Fuzzify Crisp Sensor Values 126

        5.2.3   Apply Rules 129

        5.2.4   Aggregate Active Rules and Defuzzify Output 129

  5.3   Arduino eFLL 130

        5.3.1   Example: Simple 130

        5.3.2   Example: Advanced 134

  5.4   Application 139

  5.5   Summary 146

  5.6   Problems 146

  References 147

6 Neural Networks 149

  6.1   Overview 149

  6.2   Biological Neuron 150

  6.3   Perceptron 150

        6.3.1   Training the Perceptron Model 151

        6.3.2   Single Perceptron Run Mode 160

        6.3.3   Sorting Tomatoes 161

  6.4   Multiple Perceptron Model 167

        6.4.1   Three Perceptron Run Mode 176

  6.5   Perceptron Challenges 176

  6.6   Artificial Neural Network (ANN) 177

        6.6.1   Single Neuron Model 177

        6.6.2   Single Neuron Run Mode 181

        6.6.3   Artificial Neural Networks 181

        6.6.4   ANN Convergence 197

  6.7   Deep Neural Networks–Introduction to Software Tools 198

  6.8   Application: ANN Robot Control 200

  6.9   Summary 200

  6.10 Problems 200

  References 201

Index 203

About the Author

Steven F. Barrett, Ph.D., P.E., received the BS Electronic Engineering Technology from the University of Nebraska at Omaha in 1979, the M.E.E.E. from the University of Idaho at Moscow in 1986, and the Ph.D. from The University of Texas at Austin in 1993. He was formally an active duty faculty member at the United States Air Force Academy, Colorado and is now the Vice Provost of Undergraduate Education at the University of Wyoming and Professor of Electrical and Computer Engineering. He is a member of IEEE (Life Senior) and Tau Beta Pi (chief faculty advisor). His research interests include digital and analog image processing, computer-assisted laser surgery, and embedded controller systems. He is a registered Professional Engineer in Wyoming and Colorado. He co-wrote with Dr. Daniel Pack several textbooks on microcontrollers and embedded systems. In 2004, Barrett was named "Wyoming Professor of the Year" by the Carnegie Foundation for the Advancement of Teaching and in 2008 was the recipient of the National Society of Professional Engineers (NSPE) Professional Engineers in Higher Education, Engineering Education Excellence Award.

Getting Started

1

Objectives: After reading this chapter, the reader should be able to do the following:

- Successfully download and execute a simple program using the Arduino Development Environment; and
- Describe the key features of the Arduino Development Environment.

1.1 Overview

Welcome to the world of Arduino![1] The Arduino concept of open source hardware was developed by the visionary Arduino team of Massimo Banzi, David Cuartilles, Tom Igoe, Gianluca Martino, and David Mellis in Ivrea, Italy. The team's goal was to develop a line of easy–to–use microcontroller hardware and software such that processing power would be readily available to everyone.

In this chapter we provide a brief review of the Arduino Development Environment and Arduino sketch writing. We use a top–down design approach. We begin with the "big picture" of the chapter. We then discuss the Arduino Development Environment and how it may be used to quickly develop a program (sketch) for the Arduino Nano 33 BLE Sense.

1.2 The Big Picture

Most microcontrollers are programmed with some variant of the C programming language. The C programming language provides a nice balance between the programmer's control of the microcontroller hardware and time efficiency in program (sketch) writing. As an

---

[1] This chapter is included for completeness with permission from "Arduino I: Getting Started.".

© The Author(s), under exclusive license to Springer Nature Switzerland AG 2023   1
S. F. Barrett, *Arduino V: Machine Learning*, Synthesis Lectures on Digital Circuits & Systems, https://doi.org/10.1007/978-3-031-21877-4_1

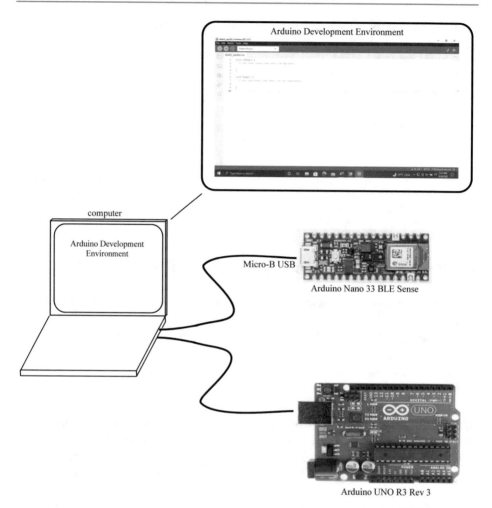

Fig. 1.1 Programming the Arduino processor board. Arduino illustrations used with permission of the Arduino Team (CC BY–NC–SA) (www.arduino.cc)

alternative, the Arduino Development Environment (ADE) provides a user–friendly interface to quickly develop a program, transform the program to machine code, and then load the machine code into the Arduino processor in several simple steps as shown in Fig. 1.1.

The first version of the Arduino Development Environment was released in August 2005. It was developed at the Interaction Design Institute in Ivrea, Italy to allow students the ability to quickly put processing power to use in a wide variety of projects. Since that time, updated versions incorporating new features, have been released on a regular basis (www.arduino.cc).

At its most fundamental level, the Arduino Development Environment is a user–friendly interface to allow one to quickly write, load, and execute code on a microcontroller. A bare-bones program need only consist of a setup() and loop() function. The Arduino Development Environment adds the other required pieces such as header files and the main program construct. The ADE is written in Java and has its origins in the Processor programming language and the Wiring Project (www.arduino.cc).

The ADE is hosted on a laptop or personal computer (PC). Once the Arduino program, referred to as a sketch, is written; it is verified and uploaded to the Arduino evaluation board.

1.3 Arduino Quickstart

To get started using an Arduino–based platform, you will need the following hardware and software:

- Arduino Nano 33 BLE Sense SoC platform,
- the appropriate interface cable from the host PC or laptop to the Arduino platform (type A to type Micro–B), and
- the Arduino Integrated Development Environment (IDE) software.

1.3.1 Quick Start Guide

The Arduino Development Environment may be downloaded from the Arduino website's at www.arduino.cc. Versions are available for Windows, Mac OS X, and Linux. Provided below is a quick start step–by–step approach to blink an onboard LED.

- Download the Arduino Development Environment from www.arduino.cc.
- Connect the Arduino processing board to the host computer via the Micro–B USB cable.
- Start the Arduino Development Environment.
- Under the Tools tab select the type of board **Board** you are using and the **Port** that it is connected to. If the board is not listed, use "Tools" − > "Manage Libraries" to access the Library Manager. From the Library Manage find and install the library to support the board. For the Arduino Nano 33 BLE Sense board the library "Arduino Mbed OS Nano Boards" is installed.
- Type the following program.

```
//********************************************************************

#define LED_PIN 13

void setup()
{
pinMode(LED_PIN, OUTPUT);        //set digital pin to output
}

void loop()
{
digitalWrite(LED_PIN, HIGH);
delay(500);                      //delay specified in ms
digitalWrite(LED_PIN, LOW);
delay(500);
}

//********************************************************************
```

- Upload and execute the program by asserting the "Upload" (right arrow) button.
- The onboard LED should blink at one second intervals.

The Arduino Nano 33 BLE Sense is equipped with a Red, Green, and Blue (RGB) LEDs. The following sketch demonstrates how to control each RGB and the Power LED. **Note: The R, G, B LEDs are asserted active low.**

```
//************************************************************************
//RGB_test
//
//Adapted from Controlling_RGB_and_Power_LED by the Arduino Team
// arduino.cc
//Demonstrates control of the RGB and Power LEDs on the NANO 33 BLE boards
//Note: The R, G, B LEDs are asserted active low.
//************************************************************************

#define RED      22              //provide pin locations of LEDs
#define GREEN    23
#define BLUE     24
#define LED_PWR 25

void setup()
{
pinMode(RED,      OUTPUT);    //initialize digital pins as output
pinMode(GREEN,    OUTPUT);
pinMode(BLUE,     OUTPUT);
pinMode(LED_PWR, OUTPUT);
}

void loop()
{
digitalWrite(RED,     HIGH); //turn LEDs off
```

```
digitalWrite(GREEN,    HIGH);
digitalWrite(BLUE,     HIGH);
digitalWrite(LED_PWR, LOW);
delay(1000);                    //delay 1s
digitalWrite(RED, LOW);         //turn RGB LEDs on in sequence
delay(1000);
digitalWrite(RED, HIGH);
delay(1000);
digitalWrite(GREEN, LOW);
delay(1000);
digitalWrite(GREEN, HIGH);
delay(1000);
digitalWrite(BLUE, LOW);
delay(1000);
digitalWrite(BLUE, HIGH);
delay(1000);
digitalWrite(LED_PWR, HIGH);
delay(1000);
}
```

```
//*********************************************************************
```

This sketch employs a function to set the R, G, and B LEDs.

```
//*********************************************************************
//RGB_test2
//
//Adapted from Controlling_RGB_and_Power_LED by the Arduino Team
// arduino.cc
//Demnstrates control of the RGB and Power LEDs on the NANO 33 BLE boards
//Note: The R, G, B LEDs are asserted active low.
//
//This sketch uses a function call to set the LED colors.
//*********************************************************************
```

```
#define RED      22             //provide pin locations of LEDs
#define GREEN    23
#define BLUE     24
#define LED_PWR 25
```

```
void setup()
{
pinMode(RED,     OUTPUT);    //intitialize digital pins as output
pinMode(GREEN,   OUTPUT);
pinMode(BLUE,    OUTPUT);
pinMode(LED_PWR, OUTPUT);
}
```

```
void loop()
{
RGB_set(LOW, HIGH,HIGH); //red
RGB_set(HIGH,LOW, HIGH); //green
RGB_set(HIGH,HIGH, LOW); //blue
```

```
RGB_set(LOW,  LOW,  HIGH);  //yellow
RGB_set(HIGH,LOW,  LOW);    //cyan
RGB_set(LOW,  HIGH,LOW);    //magenta
RGB_set(LOW,  LOW,  LOW);   //white

}

//***************************************************************

void RGB_set(bool R,  bool G,  bool B)
{
digitalWrite(RED,      HIGH);   //turn LEDs off
digitalWrite(GREEN,    HIGH);
digitalWrite(BLUE,     HIGH);
digitalWrite(LED_PWR, LOW);
delay(1000);                    //delay 1s

digitalWrite(RED,   R);         //set LEDs
digitalWrite(GREEN,G);
digitalWrite(BLUE, B);
digitalWrite(LED_PWR, HIGH);
delay(1000);                    //delay 1s

digitalWrite(RED,      HIGH);   //turn LEDs off
digitalWrite(GREEN,    HIGH);
digitalWrite(BLUE,     HIGH);
digitalWrite(LED_PWR, LOW);
delay(1000);                    //delay 1s
}

//***************************************************************
```

With the Arduino Development Environment downloaded and exercised, let's take a closer look at its features.

1.3.2 Arduino Development Environment Overview

The Arduino Development Environment is illustrated in Fig. 1.2. The ADE contains a text editor, a message area for displaying status, a text console, a tool bar of common functions, and an extensive menuing system. The ADE also provides a user–friendly interface to the Arduino processor board which allows for a quick upload of code. This is possible because the Arduino processing boards are equipped with a bootloader program.

1.3.3 Sketchbook Concept

In keeping with a hardware and software platform for students of the arts, the Arduino environment employs the concept of a sketchbook. An artist maintains their works in progress in a sketchbook. Similarly, programs are maintained within a sketchbook in the Arduino environment. Furthermore, we refer to individual programs as sketches. An individual sketch within the sketchbook may be accessed via the Sketchbook entry under the file tab.

1.3.4 Arduino Software, Libraries, and Language References

The Arduino Development Environment has a number of built–in features. Some of the features may be directly accessed via the Arduino Development Environment drop down toolbar illustrated in Fig. 1.2. Provided in Fig. 1.3 is a handy reference to show the available features. The toolbar provides a wide variety of features to compose, compile, load and execute a sketch.

1.3.5 Writing an Arduino Sketch

The basic format of the Arduino sketch consists of a "setup" and a "loop" function. The setup function is executed once at the beginning of the program. It is used to configure pins, declare variables and constants, etc. The loop function will execute sequentially step–by-

Fig. 1.2 Arduino development environment (www.arduino.cc)

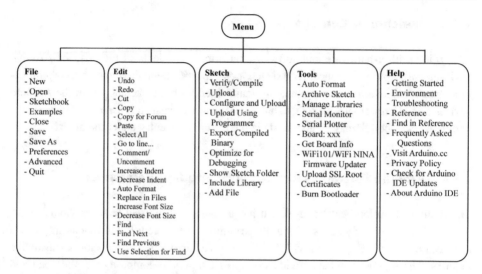

Fig. 1.3 Arduino development environment menu (www.arduino.cc)

−step. When the end of the loop function is reached it will automatically return to the first step of the loop function and execute again. This goes on continuously until the program is stopped.

```
//**********************************************************************

void setup()
  {
  //place setup code here
  }

void loop()
  {
  //main code steps are provided here
  :
  :

  }

//**********************************************************************
```

Example: Let's revisit the sketch provided earlier in the chapter.

```
//******************************************************************

#define LED_PIN 13                    //name pin 13 LED_PIN

void setup()
{
pinMode(LED_PIN, OUTPUT);             //set pin to output
}

void loop()
{
digitalWrite(LED_PIN, HIGH);          //write pin to logic high
delay(500);                           //delay specified in ms
digitalWrite(LED_PIN, LOW);           //write to logic low
delay(500);                           //delay specified in ms
}

//******************************************************************
```

In the first line the #define statement links the designator "LED_PIN" to pin 13 on the Arduino processor board. In the setup function, LED_PIN is designated as an output pin. Recall the setup function is only executed once. The program then enters the loop function that is executed sequentially step–by–step and continuously repeated. In this example, the LED_PIN is first set to logic high to illuminate the LED onboard the Arduino processing board. A 500 ms delay then occurs. The LED_PIN is then set low. A 500 ms delay then occurs. The sequence then repeats.

Even the most complicated sketches follow the basic format of the setup function followed by the loop function. To aid in the development of more complicated sketches, the Arduino Development Environment has many built–in features that may be divided into the areas of structure, variables and functions. The structure and variable features follow rules similar to the C programming language. The built–in functions consists of a set of pre–defined activities useful to the programmer. These built–in functions are summarized in Fig. 1.4.

There are many program examples available to allow the user to quickly construct a sketch. These programs are summarized in Fig. 1.5. Complete documentation for these programs is available at the Arduino homepage (www.arduino.cc). This documentation is easily accessible via the Help tab on the Arduino Development Environment toolbar. This documentation will not be repeated here. With the Arduino open source concept, users throughout the world are constantly adding new built–in features. As new features are added, they are released in future Arduino Development Environment versions. As an Arduino user, you too may add to this collection of useful tools. Throughout the remainder of the book we use both the Arduino Development Environment to program the Arduino Nano 33 BLE Sense. In the next chapter we get acquainted with the features of the Nano 33.

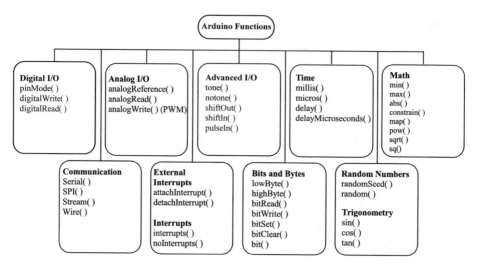

Fig. 1.4 Arduino development environment functions (www.arduino.cc)

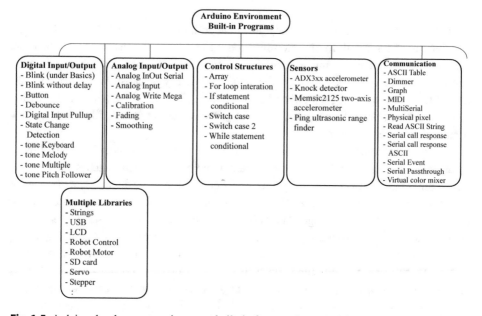

Fig. 1.5 Arduino development environment built–in features (www.arduino.cc)

1.4   Application: LED Strip

Example: LED strips may be used for motivational (fun) optical displays, games, or for instrumentation–based applications. In this example we control an LPD8806–based LED strip using the Arduino Nano 33 BLE sense. We use a three meter, 96 RGB LED strip available from Adafruit (#306) for approximately $30 USD per meter (www.adafruit.com).

The red, blue, and green component of each RGB LED is independently set using an eight–bit code. The most significant bit (MSB) is logic one followed by seven bits to set the LED intensity (0 to 127). The component values are sequentially shifted out of the Arduino 33 BLE Sense using the Serial Peripheral Interface (SPI) features as shown in Fig. 1.6a. We discuss the SPI subsystem in the next chapter.

The first component value shifted out corresponds to the LED nearest the microcontroller. Each shifted component value is latched to the corresponding R, G, and B component of the

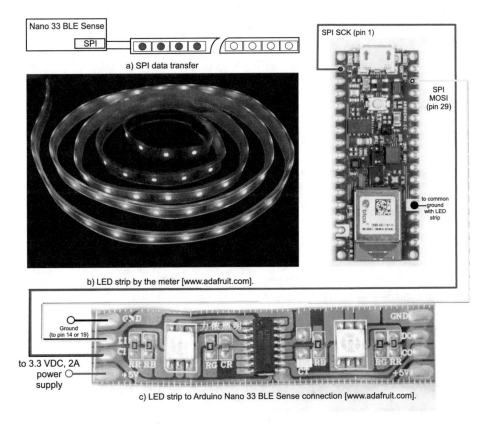

a) SPI data transfer

b) LED strip by the meter [www.adafruit.com].

c) LED strip to Arduino Nano 33 BLE Sense connection [www.adafruit.com].

Fig. 1.6 Nano 33 BLE Sense controlling an LED strip. LED strip illustration used with permission of Adafruit (www.adafruit.com). Nano 33 BLE Sense illustration used with permission of the Arduino Team (CC BY–NC–SA) (www.arduino.cc)

LED. As a new component value is received, the previous value is latched and held constant. An extra byte is required to latch the final parameter value. A zero byte $(00)_{16}$ is used to complete the data sequence and reset back to the first LED (www.adafruit.com).

Only four connections are required between the Nano 33 and the LED strip as shown in Fig. 1.6. The connections are color coded: red–power, black–ground, yellow–data, and green–clock. It is important to note the LED strip requires a supply of 3.3 VDC and a current rating of 2 amps per meter of LED strip.

In this example each RGB component is sent separately to the strip. The example illustrates how each variable in the program controls a specific aspect of the LED strip. Here are some important implementation notes:

- SPI must be configured for most significant bit (MSB) first.
- LED brightness is seven bits. Most significant bit (MSB) must be set to logic one.
- Each LED requires a separate R–G–B intensity component. The order of data is G–R–B.
- After sending data for all LEDs. A byte of (0x00) must be sent to return the strip to the first LED.
- Data stream for each LED is: 1–G6–G5–G4–G3–G2–G1–G0–1–R6–R5–R4–R3–R2–R1–R0–1–B6–B5–B4–B3–B2–B1–B0

```
//*********************************************************************
//RGB_led_strip_tutorial: illustrates different variables within
//RGB LED strip
//
//LED strip LDP8806 - available from \url{www.adafruit.com} (#306)
//
//Connections:
// - External 3.3 VDC supply, 2A per LED meter - red
// - Ground - black - include common ground with Nano BLE Sense
// - Serial Data In  - Arduino pin 29 (MOSI pin)- yellow
// - CLK  Arduino pin 1 (SCK pin)- green
//
//Variables:
// - LED_brightness - set intensity from 0 to 127
// - segment_delay  - delay between LED RGB segments
// - strip_delay    - delay between LED strip update
//
//Notes:
// - SPI must be configured for Most Significant Bit (MSB) first
// - LED brightness is seven bits.  Most Significant Bit (MSB)
//   must be set to logic one
// - Each LED requires a separate R-G-B intensity component.  The order
//   of data is G-R-B.
// - After sending data for all strip LEDs.  A byte of (0x00) must
//   be sent to return strip to first LED.
// - Data stream for each LED is:
//1-G6-G5-G4-G3-G2-G1-G0-1-R6-R5-R4-R3-R2-R1-R0-1-B6-B5-B4-B3-B2-B1-B0
//
//This example code is in the public domain.
//*********************************************************************
```

```
#include <SPI.h>

#define LED_strip_latch 0x00

const byte strip_length = 96;      //number of RGB LEDs in strip
const byte segment_delay = 10;     //delay in milliseconds
const byte strip_delay = 10;       //delay in milliseconds
unsigned char   LED_brightness;    //0 to 127
unsigned char   position;          //LED position in strip

void setup()
  {
  SPI.begin();                     //SPI support functions
  }

void loop()
  {
  SPI.beginTransaction(SPISettings(200000, MSBFIRST, SPI_MODE3));
  SPI.transfer(LED_strip_latch);   //reset to first segment
  clear_strip();                   //all strip LEDs to black
  delay(50);

  //increment the green intensity of the strip LEDs
  for(LED_brightness = 0; LED_brightness <= 60;
      LED_brightness = LED_brightness + 10)
    {
    for(position = 0; position<strip_length; position = position+1)
      {
      SPI.transfer(0x80 | LED_brightness);  //Green - MSB 1
      SPI.transfer(0x80 | 0x00);            //Red   - none
      SPI.transfer(0x80 | 0x00);            //Blue  - none
      delay(segment_delay);
      }
    SPI.transfer(LED_strip_latch);          //reset to first segment
    delay(strip_delay);
    }

  clear_strip();                            //all strip LEDs to black
  delay(50);

  //increment the red intensity of the strip LEDs
  for(LED_brightness = 0; LED_brightness <= 60;
      LED_brightness = LED_brightness + 10)
    {
    for(position = 0; position<strip_length; position = position+1)
      {
      SPI.transfer(0x80 | 0x00);            //Green - none
      SPI.transfer(0x80 | LED_brightness);  //Red   - MSB1
      SPI.transfer(0x80 | 0x00);            //Blue  - none
      delay(segment_delay);
      }
    SPI.transfer(LED_strip_latch);          //reset to first segment
```

```
    delay(strip_delay);
    }

  clear_strip();                                   //all strip LEDs to black
  delay(50);

  //increment the blue intensity of the strip LEDs
  for(LED_brightness = 0; LED_brightness <= 60;
      LED_brightness = LED_brightness + 10)
    {
    for(position = 0; position<strip_length; position = position+1)
      {
      SPI.transfer(0x80 | 0x00);                    //Green - none
      SPI.transfer(0x80 | 0x00);                    //Red   - none
      SPI.transfer(0x80 | LED_brightness);  //Blue  - MSB1
      delay(segment_delay);
      }
    SPI.transfer(LED_strip_latch);                //reset to first segment
    delay(strip_delay);
    }

  clear_strip();                                   //all strip LEDs to black
  SPI.endTransaction();
  delay(50);
  }

//******************************************************************

void clear_strip(void)
{
  //clear strip
  for(position = 0; position<strip_length; position = position+1)
      {
      SPI.transfer(0x80 | 0x00);                    //Green - none
      SPI.transfer(0x80 | 0x00);                    //Red - none
      SPI.transfer(0x80 | 0x00);                    //Blue  - none
      }
  SPI.transfer(LED_strip_latch);                //Latch with zero
  delay(200);                                       //clear delay
}

//******************************************************************
```

1.5   Summary

The goal of this chapter is to provide an introduction and tutorial on the Arduino IDE. We used a top–down design approach. We began with the "big picture" of the chapter followed by an overview of the Arduino Development Environment.

1.6   Problems

1. Describe the steps in writing a sketch and executing it on an Arduino processing board.
2. What is the serial monitor feature used for in the Arduino Development Environment?
3. Describe what variables are required and returned and the basic function of the following built–in Arduino functions: Blink, Analog Input.
4. What is meant by the term open source?
5. The RGB LEDs onboard the Nano 33 BLE are active low. What does this mean?
6. Be creative! Modify the sketch controlling the strip LEDs to generate a different pattern. Have fun!

References

1. Arduino homepage, www.arduino.cc
2. *Arduino Nano 33 BLE Sense*, ABX00031, January 5, 2022. www.arduino.cc

Arduino Nano 33 BLE Sense

2

Objectives: After reading this chapter, the reader should be able to do the following:

- Name and describe the different subsystem peripherals onboard the Nordic Semiconductor nRF52840 processor;
- Name and describe the different features aboard the Arduino Nano 33 BLE Sense board;
- In your own words describe the background theory of operation for subsystems onboard the nRF52840 processor and Arduino Nano 33 BLE Sense board; and
- Use the Arduino IDE to program and execute sketches for subsystems onboard the nRF52840 processor and Arduino Nano 33 BLE Sense board.

2.1 Overview

In this chapter we explore the Arduino Nano 33 BLE Sense SoC board and its nRF52840 processor. We begin with an overview of the Arduino Nano 33 BLE Sense SoC board features. We then examine the powerful and well–equipped nRF52840 processor and its associated peripheral subsystems. We then investigate the multiple peripherals onboard the Nano 33 BLE Sense SoC board. For all peripherals we provide a brief theory of operation, feature overview, and examples. We conclude the chapter with an extended example featuring a Bluetooth BLE application–a greenhouse monitoring system.[1]

---

[1] Portions of the theory provided in the chapter was adapted with permission from "Arduino I: Getting Started, S. Barrett, Morgan & Claypool Publishers, 2020.

S. F. Barrett, *Arduino V: Machine Learning*, Synthesis Lectures on Digital Circuits & Systems, https://doi.org/10.1007/978-3-031-21877-4_2

2.2 Arduino Nano 33 BLE Sense SoC Board

The Arduino Nano 33 BLE Sense SoC board is illustrated in Fig. 2.1. **The Nano 33 is a 3.3 VDC processor.** Working clockwise from the left, the board is equipped with a Micro–B USB connector to allow programming the processor from a host personal computer (PC) or laptop.

The board is equipped with a series of LEDs including the Power LED (L2), the Red, Green, Blue (RGB) LEDs (DL3), and the Built–In LED (L1). As we experienced in Chap. 1, these LEDs are accessible via designated pins.

The Arduino Nano 33 BLE Sense SoC board is equipped with a rich complement of sensors including the (ABX00031):

- Nine axis inertial measurement unit (IMU) (LSM9DS1),
- Barometer and temperature sensor (LPS22HB),
- Relative humidity sensor (HTS221),
- Digital proximity, ambient light, RGB, and gesture sensor (APDS–9960),
- Digital microphone (MP34DT05),

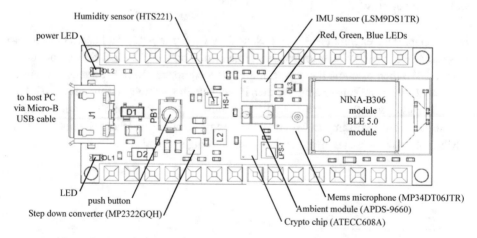

Fig. 2.1 Arduino Nano 33 BLE Sense SoC board layout. (Figure adapted and used with permission of Arduino Team (CC BY–NC–SA)(www.arduino.cc).)

- Crypto Chip (ATEC608A), and the
- DC–DC converter (MPM3610).

Onboard the Nano 33 board is the NINA B306 Module. It contains the Nordic Semi-conductor nRF52840 processor. This is a 32–bit processor, operating at 64 MHz, with an ARM Cortex–M4F architecture equipped with a floating point unit (FPU). The processor is equipped with 1 MB of flash memory and 256 kB of Random Access Memory (RAM) (nRF52840).

The Nano 33 BLE provides a tremendous amount of computing power in a very small footprint. The processing power coupled with its peripherals and sensor package provides a microcontroller ideally suited for AI and ML applications.

The B306 module is also equipped with a large complement of resident peripherals including (ABX00031):

- Serial communication subsystems including Universal Serial Bus (USB), Universal Asynchronous Receiver Transmitter (UART), Serial Peripheral Interface (SPI), Quad SPI (QSPI), Two Wire Interface (TWI).
- A 12–bit resolution, 200 ksps (kilo samples per second) analog–to–digital converter (ADC),
- A 128 bit Advanced Encryption Standard (AES) co–processor,
- Bluetooth and Zigbee radio,
- Near Field Communication (NFC) features,
- Direct Memory Access (DMA) capability,
- An ARM CC310 Crytocell cryptographic accelerator, and
- Pulse Width Modulation (PWM) channels.

Access to these subsystems are via the Arduino Nano 33 BLE Sense SoC board pins as shown in Fig. 2.2.

2.3   Arduino Nano 33 BLE Sense Features

With the brief overview of the Arduino Nano 33 BLE Sense board complete, we take an in–depth look at selected features and subsystems. As described in the previous section, we partition features related to the NINA B306 Module from the Arduino Nano 33 BLE Sense as shown in Fig. 2.3. We begin with an exploration of NINA B306 Module subsystems followed by Arduino Nano 33 BLE Sense subsystems. For each subsystem we provide related technical information and examples where appropriate.

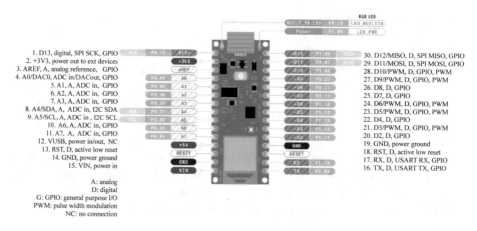

1. D13, digital, SPI SCK, GPIO
2. +3V3, power out to ext devices
3. AREF, A, analog reference, GPIO
4. A0/DAC0, ADC in/DACout, GPIO
   5. A1, A, ADC in, GPIO
   6. A2, A, ADC in, GPIO
   7. A3, A, ADC in, GPIO
8. A4/SDA, A, ADC in, I2C SDA
9. A5/SCL, A, ADC in , I2C SCL
   10. A6, A, ADC in, GPIO
   11. A7, A, ADC in, GPIO
   12. VUSB, power in/out, NC
   13. RST, D, active low reset
   14. GND, power ground
   15. VIN, power in

30. D12/MISO, D, SPI MISO, GPIO
29. D11/MOSI, D, SPI MOSI, GPIO
28. D10/PWM, D, GPIO, PWM
27. D9/PWM, D, GPIO, PWM
26. D8, D, GPIO
25. D7, D, GPIO
24. D6/PWM, D, GPIO, PWM
23. D5/PWM, D, GPIO, PWM
22. D4, D, GPIO
21. D3/PWM, D, GPIO, PWM
20. D2, D, GPIO
19. GND, power ground
18. RST, D, active low reset
17. RX, D, USART RX, GPIO
16. TX, D, USART TX, GPIO

A: analog
D: digital
G: GPIO: general purpose I/O
PWM: pulse width modulation
NC: no connection

Fig. 2.2 Arduino Nano 33 BLE Sense SoC board pin out. (Figure adapted and used with permission of Arduino Team (CC BY–NC–SA)(www.arduino.cc).)

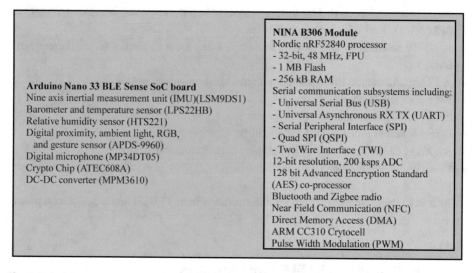

Arduino Nano 33 BLE Sense SoC board
Nine axis inertial measurement unit (IMU)(LSM9DS1)
Barometer and temperature sensor (LPS22HB)
Relative humidity sensor (HTS221)
Digital proximity, ambient light, RGB,
   and gesture sensor (APDS-9960)
Digital microphone (MP34DT05)
Crypto Chip (ATEC608A)
DC-DC converter (MPM3610)

NINA B306 Module
Nordic nRF52840 processor
- 32-bit, 48 MHz, FPU
- 1 MB Flash
- 256 kB RAM
Serial communication subsystems including:
- Universal Serial Bus (USB)
- Universal Asynchronous RX TX (UART)
- Serial Peripheral Interface (SPI)
- Quad SPI (QSPI)
- Two Wire Interface (TWI)
12-bit resolution, 200 ksps ADC
128 bit Advanced Encryption Standard
(AES) co-processor
Bluetooth and Zigbee radio
Near Field Communication (NFC)
Direct Memory Access (DMA)
ARM CC310 Crytocell
Pulse Width Modulation (PWM)

Fig. 2.3 Arduino Nano 33 BLE Sense subsystems (nRF52840)

2.4   NINA B306 Module Subsystems

In this section we explore the peripherals contained within the NINA B306 Module. The B306 is equipped with the Nordic Semiconductor nRF52840 processor. This is a 32–bit processor, operating at 64 MHz, with an ARM Cortex–M4F architecture equipped with a floating point unit (FPU). In addition, it is equipped with specialized instructions to aid in efficient AI and ML program execution including (nRF52840):

- Digital Signal Processing (DSP) instruction set,
- Direct Memory Access (DMA) for efficient CPU and peripheral access to memory,
- Hardware based arithmetic divider to accelerate this operation, and
- Multiple and accumulate (MAC) instructions requiring a single clock cycle for execution.

The 32–bit processor architecture allows a wide range of integer and floating point (real number) processing capability. The processor is equipped with a floating point unit to enhance processing. The processor's 48 MHz clock speed provides enhanced capability over other Arduino products. For example, the very capable Arduino UNO R3 processor hosting the Microchip ATmega328 operates at 16 MHz.

2.4.1   B306 Module Memory

The B306 is equipped with two main memory sections: flash electrically erasable programmable read only memory (EEPROM) and static random access memory (SRAM). The processor is equipped with 1 MB (megabyte) of flash memory and 256 KB (kilobyte) of Random Access Memory (RAM) (nRF52840). We discuss each memory component in turn.

2.4.1.1 B306 Programmable Flash EEPROM

Bulk programmable flash EEPROM is used to store programs. It can be erased and programmed as a single unit. Also, should a program require a large table of constants, it may be included as a global variable within a program and programmed into flash EEPROM with the rest of the program. Flash EEPROM is nonvolatile meaning memory contents are retained even when microcontroller power is lost. The B306 is equipped with 1 MB bytes of onboard reprogrammable flash memory.

2.4.1.2 B306 Static Random Access Memory (SRAM)

Static RAM memory is volatile. That is, if the microcontroller loses power, the contents of SRAM memory are lost. It can be written to and read from during program execution. The B306 is equipped with 256 KB of SRAM. A small portion of the SRAM is set aside for the general–purpose registers used by the processor and also for the input/output and peripheral subsystems aboard the microcontroller. During program execution, RAM is used to store program instructions, global variables, support dynamic memory allocation of variables, and to provide a location for the stack. The large SRAM component is especially useful in certain AI and ML applications.

2.5   NINA B306 Module Peripherals

In this section, we provide a brief overview of the internal peripherals of the B306 module. It should be emphasized that these features are the internal systems contained within the confines of the microcontroller chip. These built–in peripherals allow complex and sophisticated tasks to be accomplished by the microcontroller.

2.5.1   Pulse Width Modulation (PWM) Channels

The Nano 33 BLE Sense is equipped with five pulse width modulation (PWM) channels. PWM output can be provided on digital pins 1–13 and analog pins A0–A7. We limit use to pins 21, 23, 24, 27, and 28 as shown in the pinout diagram at Fig. 2.2. The Nano 33 BLE sense baseline frequency of the PWM signal is set at 500 Hz.

A pulse width modulated or PWM signal is characterized by a fixed frequency and a varying duty cycle. Duty cycle is the percentage of time a repetitive signal is logic high during the signal period. It may be formally expressed as:

$$duty\ cycle[\%] \ = \ (on\ time/period) \ \times \ (100\%)$$

To generate a PWM signal within the Arduino IDE, the AnalogWrite() function is used. The function is called with the desired PWM output pin (pin) and the desired PWM duty cycle value. The duty cycle value is specified in the range of 0% (0) to 100% (255).

```
analogWrite(pin, value)
```

Example: In this example the intensity of an LED, connected to Nano 33 BLE Sense pin 21 (digital D3), is slowly increased and then decreased using PWM techniques. Note the use of a 220 Ohm resistor in series with the LED. We discuss interface techniques in Chap. 3.

```
//*******************************************************
//pwm_LED_test: the intensity of an LED, connected to
//Nano 33 BLE Sense pin 21, is slowly increased and then
//decreased using PWM techniques.
//
//Notes:
//- Provide a 220 Ohm resistor in series with the LED
//   to common ground with the Nano 33.
//- Nano 33 BLE Sense pin 21 is specified as digital D3
//   in the Arduino IDE sketch.
//*******************************************************

int ext_red_LED_pin = 3;      //LED on D3 (physical pin 21)
```

```
void setup()
{
pinMode(ext_red_LED_pin, OUTPUT);   //sets the pin as output
}

void loop()
{
int LED_int = 0;
                              //increase intensity
for(LED_int=0; LED_int<=255; LED_int=LED_int+5)
  {                           //0% (0) to 100% (255)
  analogWrite(ext_red_LED_pin, LED_int);
  delay(100);                 //delay 100 ms
  }

LED_int = 255;
                              //decrease intensity
for(LED_int=255; LED_int>=0; LED_int=LED_int-5)
  {                           //100% (255) to 0% (0)
  analogWrite(ext_red_LED_pin, LED_int);
  delay(100);                 //delay 100 ms
  }
delay(100);                   //delay 100 ms
}

//************************************************************
```

PWM signals are used in a wide variety of applications including controlling the position of a servo motor and controlling the speed of a DC motor. We explore these applications in the next chapter.

2.5.2   Serial Communications

The Arduino Nano 33 BLE Sense is equipped with a variety of different serial communication subsystems including the Universal Synchronous and Asynchronous Serial Receiver and Transmitter (USART), the serial peripheral interface (SPI), and the Two–wire Serial Interface (TWI). What these systems have in common is the serial transmission of data. In a serial communications transmission, serial data is sent a single bit at a time from transmitter to receiver. The serial communication subsystems are typically used to add and communicate with additional peripheral devices.

2.5.2.1 USART
The serial USART may be used for full duplex (two way) communication between a receiver and transmitter. This is accomplished by equipping the Nano 33 with independent hardware

for the transmitter and receiver. The USART is typically used for asynchronous communication. That is, there is not a common clock between the transmitter and receiver to keep them synchronized with one another. To maintain synchronization between the transmitter and receiver, framing start and stop bits are used at the beginning and end of each data byte in a transmission sequence.

The Nano 33 USART is quite flexible. It has the capability to be set to different data transmission rates known as the Baud (bits per second) rate. The USART may also be set for several data bit widths and different BAUD rates. Furthermore, it is equipped with a hardware generated parity bit (even or odd) and parity check hardware at the receiver. A single parity bit allows for the detection of a single bit error within a byte of data.

Example: In this example we equip the Nano 33 with a liquid crystal display (LCD). An LCD is an output device to display text information as shown in Fig. 2.5. LCDs come in a wide variety of configurations including multi–character, multi–line format. A 16 x 2 LCD format is common. That is, it has the capability of displaying two lines of 16 characters each.

Characters are sent to the LCD via the American Standard Code for Information Interchange (ASCII) format a single character or control command at a time. ASCII is a standardized, seven bit method of encoding alphanumeric data. It has been in use for many decades, so some of the characters and actions listed in the ASCII table are not in common use today. However, ASCII is still the most common method of encoding alphanumeric data. The ASCII code is shown in Fig. 2.4. We illustrate the use of the table with an example. The capital letter "G" is encoded in ASCII as 0x47. The "0x" symbol indicates the hexadecimal number representation.

Unicode is the international counterpart of ASCII. It provides a standardized 16–bit encoding format for the written languages of the world. ASCII is a subset of Unicode. The interested reader is referred to the Unicode home page website at: www.unicode.org for additional information on this standardized encoding format.

LCDs are configured for either a serial or parallel interface to the host microcontroller. For a parallel configured LCD, an eight bit data path and two control lines are required between the microcontroller. Many parallel configured LCDs may also be configured for a four bit data path thus saving several precious microcontroller pins. A small microcontroller mounted to the back panel of the LCD translates the ASCII data characters and control signals to properly display the characters. Several manufacturers provide 3.3 VDC compatible displays.

To conserve precious, limited microcontroller input/output pins a serial configured LCD may be used. A serial LCD reduces the number of required microcontroller pins for interface, from ten down to one. Display data and control information is sent to the LCD via an asynchronous USART serial communication link (8 bit, 1 stop bit, no parity, 9600 Baud).

In this example a Sparkfun LCD–16397, 3.3 VDC, serial, 16 by 2 character LCD display is connected to the Nano 33 BLE Sense. Communication between the Nano 33 and the LCD is accomplished by a single 9600 bits per second (BAUD) connection using the onboard USART.

	Most significant digit							
	0x0_	0x1_	0x2_	0x3_	0x4_	0x5_	0x6_	0x7_
0x_0	NUL	DLE	SP	0	@	P	`	p
0x_1	SOH	DC1	!	1	A	Q	a	q
0x_2	STX	DC2	"	2	B	R	b	r
0x_3	ETX	DC3	#	3	C	S	c	s
0x_4	EOT	DC4	$	4	D	T	d	t
0x_5	ENQ	NAK	%	5	E	U	e	u
0x_6	ACK	SYN	&	6	F	V	f	v
0x_7	BEL	ETB	'	7	G	W	g	w
0x_8	BS	CAN	(	8	H	X	h	x
0x_9	HT	EM	)	9	I	Y	i	y
0x_A	LF	SUB	*	:	J	Z	j	z
0x_B	VT	ESC	+	;	K	[	k	{
0x_C	FF	FS	'	<	L	\	l		
0x_D	CR	GS	-	=	M	]	m	}
0x_E	SO	RS	.	>	N	^	n	~
0x_F	SI	US	/	?	O	_	o	DEL

(Least significant digit labels the leftmost column rows)

Fig. 2.4 ASCII Code. The ASCII code is used to encode alphanumeric characters. The "0x" indicates hexadecimal notation in the C programming language

Fig. 2.5 LCD display with serial interface

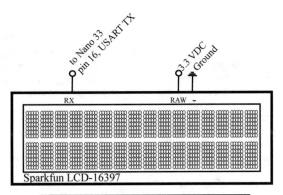

Line	Character Position (n)
1	0-15
2	64-79

Note: character position is specifed as $0X80 + n$

Command Code	Command
0x01	Clear Display
0x14	Cursor one space right
0x10	Cursor one space left
0x80 + n	Cursor to position

Note: precede command with 0xFE (254_{10})

In the Application section of Chap. 3, we configure a small robot to navigate autonomously through a maze. The robot will display maze wall status on an LCD. In this sample sketch we use simulated maze wall data to display on the LCD. Note the USART is designated "Serial1" in the sketch.

```
//****************************************************************************
//LCD_example
//
//Sparkfun LCD-16397, 3.3 VDC, 16x2 character display
//- Nano 33 BLE Sense, USART TX pin 16 is connected to LCD USART RX pin
//- Provide 3.3 VDC power to the LCD
//****************************************************************************

int left_IR_sensor_value =   0;       //variable for left IR sensor
int center_IR_sensor_value = 0;       //variable for center IR sensor
int right_IR_sensor_value  = 0;       //variable for right IR sensor

void setup()
{
Serial1.begin(9600);                  //Baud rate: 9600 Baud
delay(500);                           //Delay for display
}

void loop()
{
//read analog output from IR sensors Â– simulated maze wall data
left_IR_sensor_value   = left_IR_sensor_value   + 1;
center_IR_sensor_value = center_IR_sensor_value + 2;
right_IR_sensor_value  = right_IR_sensor_value  + 3;

//Clear LCD
//Cursor to line one, character one
Serial1.write(254);                   //Command prefix
Serial1.write(128);                   //Command

//clear display
Serial1.write("                ");
Serial1.write("                ");

//Cursor to line one, character one
Serial1.write(254);                   //Command prefix
Serial1.write(128);                   //Command
Serial1.write("Left  Ctr  Right");
delay(50);

Serial1.write(254);                   //Command to LCD
delay(5);
Serial1.write(192);                   //Cursor line 2, position 1
delay(5);
Serial1.print(left_IR_sensor_value);
delay(5);
Serial1.write(254);                   //Command to LCD
```

```
delay(5);
Serial1.write(198);                      //Cursor line 2, position 8
delay(5);
Serial1.print(center_IR_sensor_value);
delay(5);
Serial1.write(254);                      //Command to LCD
delay(5);
Serial1.write(203);                      //Cursor line 2, position 13
delay(5);
Serial1.print(right_IR_sensor_value);
delay(5);
delay(500);
}
```

//**

2.5.2.2 Serial Peripheral Interface (SPI)

The Nano 33 BLE Sense Serial Peripheral Interface or SPI provides for two–way serial communication between a transmitter and a receiver. In the SPI system, the transmitter and receiver share a common clock source. This requires an additional clock line between the transmitter and receiver but allows for higher data transmission rates as compared to the USART.

The SPI system allows for fast and efficient data exchange between microcontrollers or peripheral devices. There are many SPI compatible external systems available to extend the features of the microcontroller. For example, a liquid crystal display or a digital–to–analog converter could be added to the microcontroller using the SPI system.

SPI Operation The SPI may be viewed as a synchronous 16–bit shift register with an 8–bit half residing in the transmitter and the other 8–bit half residing in the receiver as shown in Fig. 2.6. The transmitter is designated the master since it is providing the synchronizing clock source between the transmitter and the receiver. The receiver is designated as the slave. A slave is chosen for reception by taking its Slave Select ($\overline{SS}$) line low. When the $\overline{SS}$ line is taken low, the slave's shifting capability is enabled.

SPI transmission is initiated by loading a data byte into the master configured SPI data register. At that time, the SPI clock generator provides clock pulses to the master and also to the slave via the SCK pin. A single bit is shifted out of the master designated shift register on the Master Out Slave In (MOSI) microcontroller pin on every SCK pulse.

The data is received at the MOSI pin of the slave designated device. At the same time, a single bit is shifted out of the Master In Slave Out (MISO) pin of the slave device and into the MISO pin of the master device.

After eight master SCK clock pulses, a byte of data has been exchanged between the master and slave designated SPI devices.

The SPI associated pins on the Arduino Nano 33 BLE Sense include:

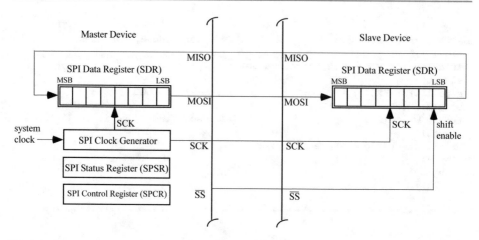

Fig. 2.6 SPI overview

- Pin 29, SPI MOSI also referred to as Computer Out Peripheral In (COPI),
- Pin 30, SPI MISO also referred to as Computer In Peripheral Out (CIPO),
- Pin 1, SPI SCK

To configure the Arduino Nano 33 BLE Sense for SPI operation, the following Arduino IDE commands are used:

- SPI.begin() is called from within setup()
- SPI.beginTransaction(SPISettings(SPIspeed, SPIbitdirection, SPImode));
- SPI.transfer(data)
- SPI.endTransaction();

In Chap. 1 we used the SPI system to send RGB data to individual LEDs within an SPI compatible LED strip.

Example: In this example we configure the Sparkfun LCD–16397. The required connections, shown in Fig. 2.7, between the Nano 33 and the LCD include:

- Nano 33 SCK pin 1 to SCK on LCD
- Nano 33 SPI MOSI pin 29 to SDI (Serial Data In) on LCD
- Ground LCD chip select (\CS)

Provided below is an Arduino sketch to communicate with LCD–16397 using the SPI system. Several items of interest regarding the sketch:

Fig. 2.7 LCD display with the
serial peripheral interface

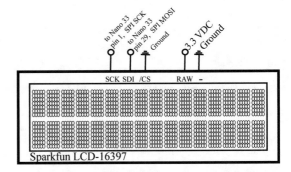

Sparkfun LCD-16397

Line	Character Position (n)
1	0-15
2	64-79

Note: character position is specifed
as 0X80 + n

Command Code	Command
0x01	Clear Display
0x14	Cursor one space right
0x10	Cursor one space left
0x80 + n	Cursor to position

Note: precede command with 0xFE (254_{10})

- SparkFun documentation for the LCD provide the following LCD SPI settings: speed-100000, MSBFIRST, SPI_MODE0.
- Note the use of function **transmit_int_via_SPI** to send a three digit integer value to the LCD. The function isolates each digit and sends its ASCII equivalent to the LCD.

```
//*********************************************************************
//LCD_SPI_example
//
//The robot is equipped with a Sparkfun LCD-16397
// - Nano 33 SCK pin 1 to SCK on LCD
// - Nano 33 SPI MOSI pin 29 to SDI (Serial Data In) on LCD
// - Ground LCD chip select (\CS)
//- Provide 3.3 VDC power to the LCD
//*********************************************************************

#include <SPI.h>

int left_IR_sensor_value =   0;      //variable for left IR sensor
int center_IR_sensor_value = 0;      //variable for center IR sensor
int right_IR_sensor_value  = 0;      //variable for right IR sensor

void setup()
{
```

```
SPI.begin();
delay(500);                             //Delay for display
}

void loop()
{
SPI.beginTransaction(SPISettings(100000, MSBFIRST, SPI_MODE0));

//read analog output from IR sensors
left_IR_sensor_value   = left_IR_sensor_value   + 1;
center_IR_sensor_value = center_IR_sensor_value + 2;
right_IR_sensor_value  = right_IR_sensor_value  + 3;

//Clear LCD
//Cursor to line one, character one
SPI.transfer(254);                      //Command prefix
SPI.transfer(128);                      //Command

//clear display
spiSendString("                ");
spiSendString("                ");

//Cursor to line one, character one
SPI.transfer(254);                      //Command prefix
SPI.transfer(128);                      //Command
spiSendString("Left  Ctr  Right");
delay(50);

SPI.transfer(254);                      //Command to LCD
delay(5);
SPI.transfer(192);                      //Cursor line 2, position 1
delay(5);
transmit_int_via_SPI(left_IR_sensor_value);
delay(5);
SPI.transfer(254);                      //Command to LCD
delay(5);
SPI.transfer(198);                      //Cursor line 2, position 8
delay(5);
transmit_int_via_SPI(center_IR_sensor_value);
delay(5);
 SPI.transfer(254);                     //Command to LCD
delay(5);
SPI.transfer(203);                      //Cursor line 2, position 13
delay(5);
transmit_int_via_SPI(right_IR_sensor_value);

SPI.endTransaction();
delay(500);
}

//******************************************************************

void spiSendString(char* data)
{
for(byte x = 0; data[x] != '\0'; x++)    //send chars to end of string
  {
```

```
  SPI.transfer(data[x]);
  }
}

//*****************************************************************

void transmit_int_via_SPI(unsigned int num_to_convert)
{
unsigned int hundreds_place, tens_place, ones_place;
char         hundreds_place_char, tens_place_char, ones_place_char;

hundreds_place = (unsigned int)(num_to_convert/100);
hundreds_place_char = (char)(hundreds_place + 48);
SPI.transfer(hundreds_place_char);
delay(5);

tens_place = (unsigned int)((num_to_convert-(hundreds_place * 100))/10);
tens_place_char = (char)(tens_place + 48);
SPI.transfer(tens_place_char);
delay(5);

ones_place=(unsigned int)(num_to_convert-(hundreds_place*100)-(tens_place*10));
ones_place_char = (char)(ones_place + 48);
SPI.transfer(ones_place_char);
delay(5);
}

//*****************************************************************
```

2.5.2.3 Inter–Integrated Circuit (I2C)

The I2C subsystem allows the system designer to network related devices (microcontrollers, transducers, displays, memory storage, etc.) together into a system using a two–wire interconnecting scheme. The I2C allows a maximum of 128 devices to be interconnected. Each device has its own unique address and may both transmit and receive over the two–wire bus at frequencies up to 400 kHz. This allows the device to freely exchange information with other devices in the network within a small area. Devices within the small area network are connected by two wires to share data (SDA) and a common clock (SCL).

```
//*****************************************************************
//LCD_I2C_example
//
//The robot is equipped with a Sparkfun LCD-16397
// - Nano 33 I2C SDA pin 8 to LCD DA pin
// - Nano 33 I2C CLK pin 9 to LCD CL
// - Ground LCD chip select (\CS)
//- Provide 3.3 VDC power to the LCD
//- LCD default I2C address: 0x72
//*****************************************************************

#include <Wire.h>

#define LCD_I2C_addr 0x72 //default address of LCD
```

```
int cycles = 0;

int left_IR_sensor_value =   0;        //variable for left IR sensor
int center_IR_sensor_value = 0;        //variable for center IR sensor
int right_IR_sensor_value  = 0;        //variable for right IR sensor


void setup()
{
Wire.begin();                          //Join I2C bus - master mode
Wire.beginTransmission(LCD_I2C_addr);
Wire.write('|');                       //LCD setting mode
Wire.write('-');                       //clear display command
Wire.endTransmission();
}

void loop()
{
//read analog output from IR sensors
left_IR_sensor_value   = left_IR_sensor_value   + 1;
center_IR_sensor_value = center_IR_sensor_value + 2;
right_IR_sensor_value  = right_IR_sensor_value  + 3;

I2CSendValue(left_IR_sensor_value, center_IR_sensor_value, right_IR_sensor_value);
delay(500);                            //delay
}

//*********************************************************************

void I2CSendValue(int value1, int value2, int value3)
{
Wire.beginTransmission(LCD_I2C_addr);     //transmit to LCD
Wire.write('|');                          //LCD setting mode
Wire.write('-');                          //clear display command

Wire.print("Left  Ctr  Right");
Wire.print(value1);
Wire.print("    ");

Wire.print(value2);
Wire.print("   ");

Wire.print(value3);
Wire.print("  ");

Wire.endTransmission();                   //Stop I2C transmission
}
//*********************************************************************
```

2.5.2.4 Analog to Digital Converter–ADC

The goal of the ADC process is to accurately represent analog signals as digital signals. Toward this end, three signal processing procedures, sampling, quantization, and encoding must be combined together.

Before the ADC process takes place, we first need to convert a physical signal into an electrical signal with the help of a transducer. A transducer is an electrical and/or mechanical system that converts physical signals into electrical signals or electrical signals to physical signals.

Depending on the purpose, we categorize a transducer as an input transducer or an output transducer. If the conversion is from physical to electrical, we call it an input transducer. For example, a temperature sensor is considered an input transducer. The output transducer converts electrical signals to physical signals. For example, an LCD or a motor would be considered an output transducer.

It is important to carefully design the interface between transducers and the microcontroller to insure proper operation. A poorly designed interface could result in improper embedded system operation or failure. Specific input and output transducer interface techniques are discussed in Chap. 3.

The Arduino Nano 33 BLE Sense is equipped with an eight–channel, 12–bit resolution, 200 kilo samples per second (ksps) analog to digital converter (ADC) subsystem. The ADC converts an analog signal from the outside world into a binary representation suitable for use by the microcontroller. The 12–bit resolution means that an analog voltage between 0 and 3.3V will be encoded into one of 4096 binary representations between $(000)_{16}$ and $(FFF)_{16}$. This provides the Nano 33 with a voltage resolution of approximately 0.81 mV.

An analog channel is read using the "analogRead" function. This function uses a default value of 10–bit ADC resolution. The resolution of the ADC process can be adjusted using the "analogReadResolution(num_bits)" function. The desired level of resolution is specified using the "num_bits" variable.

The Nano 33 has eight analog to digital conversion channels:

- Channel A0, pin 4
- Channel A1, pin 5
- Channel A2, pin 6
- Channel A3, pin 7
- Channel A4, pin 8 (also I2C pin SDA)
- Channel A5, pin 9 (also I2C pin SCL)
- Channel A6, pin 10
- Channel A7, pin 11

Example: In this example we measure the voltage from a variable power supply. Note the line in code to convert the ADC reading from 0 to 1023 to an analog voltage: `A0_voltage = (analog_reading_A0 * 3.3)/1024`. The circuit configuration is shown in Fig. 2.8.

```
//******************************************************************
//LCD_USART_ADC_example
//
//The robot is equipped with a Sparkfun LCD-16397
//- Nano 33 BLE Sense, USART TX pin 16 is connected to LCD USART RX pin
//- Provide 3.3 VDC power to the LCD
//- Nano 33 BLE Sense Analog A0 at pin 4
//******************************************************************

int analog_reading_A0;
float A0_voltage;

void setup()
{
Serial1.begin(9600);                     //Baud rate: 9600 Baud
delay(500);                              //Delay for display
}

void loop()
{
analog_reading_A0 = analogRead(A0);   //read A0
                                      //convert to voltage
A0_voltage = (analog_reading_A0 * 3.3)/1024;
//Clear LCD
//Cursor to line one, character one
Serial1.write(254);                      //Command prefix
Serial1.write(128);                      //Command

//clear display
Serial1.write("                ");
Serial1.write("                ");

//Cursor to line one, character one
Serial1.write(254);                       //Command prefix
Serial1.write(128);                       //Command
Serial1.write("Voltage:        ");
delay(50);

Serial1.write(254);                       //Command to LCD
delay(5);
Serial1.write(192);                       //Cursor line 2, position 1
delay(5);
Serial1.print(analog_reading_A0);
Serial1.write("  ");
Serial1.print(A0_voltage);
Serial1.write("V");

delay(500);
}

//******************************************************************
```

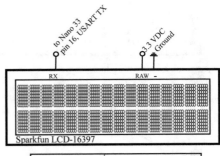

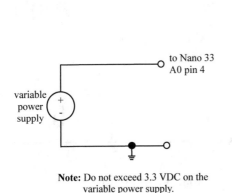

Line	Character Position (n)
1	0-15
2	64-79

Note: character position is specifed as 0X80 + n

Command Code	Command
0x01	Clear Display
0x14	Cursor one space right
0x10	Cursor one space left
0x80 + n	Cursor to position

Note: precede command with 0xFE (254_{10})

Fig. 2.8 ADC test with LCD display

2.5.3   Bluetooth Low Energy (BLE)

The Arduino Nano 33 BLE Sense is equipped with Bluetooth features. The Classic form of Bluetooth was designed to provide a wireless replacement for the common RS–232 serial connection standard. The Arduino Nano 33 BLE sense is also equipped with Bluetooth Low Energy (BLE) features. It is important to note that Bluetooth Classic and BLE features are not compatible with one another. We explore Bluetooth Classic in "Arduino III: Internet of Things."[2] We concentrate on BLE features here.

Bluetooth BLE provides for low transmit power (10 mW), short (maximum 100 m) range RF connections to replace wires. It uses the crowded Industrial, Scientific, and Medical (ISM) frequency band from 2.40 to approximately 2.50 GHz. The BLE band is divided into 40 different, 2 MHz channels as shown in Fig. 2.9. BT BLE employs an interesting frequency hopping technique to communicate. Data for transmission is divided into packets at data rates from 125 to 2 Mb/s. The device transmits a packet of data at the first carrier frequency. It then hops to a different carrier frequency for the next packet and so on until the entire message is transmitted as shown in Fig. 2.9b). Formally the BT BLE modulation technique is called Direct Sequence Spread Spectrum (DSSS) (www.bluetooth.com).

BLE uses the Generic Attribute (GATT) Profile to establish two different primary roles for a BLE connection:

---

[2] "Arduino III: Internet of Things," S.F. Barrett, Morgan and Claypool Publishers, 2021.

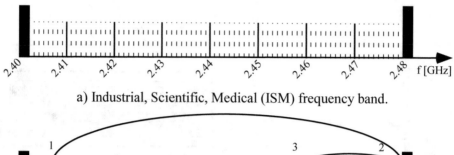

a) Industrial, Scientific, Medical (ISM) frequency band.

b) Frequency Hopping Spread Spectrum (FHSS) [R and S].

Fig. 2.9 Bluetooth BLE communication concepts

- The peripheral or server role provides bulletin board features where data is posted for reading.
- The central or client role can read and interact with the posted data.

In Fig. 2.10 we use an Arduino Nano 33 BLE Sense in a peripheral server role to collect important greenhouse information such as external temperature, internal temperature, humidity, and soil moisture content. The greenhouse related data is collected and organized into a BLE service. The service related data is provided as BLE configured characteristics. To allow ease of access to the information from an external central client device, the BLE service and characteristics are each assigned a universally unique identifier (UUID) (www. bluetooth.com). If we were to expand the features of the project with additional services, we could group them into a profile.

There are a number of 16 bit pre–assigned UUIDs. The UUIDs represent different manufacturers and technology companies employing Bluetooth–based technologies. Also, UUIDs have been pre–assigned to common Bluetooth features and common pre–assigned data types (e.g. temperature, pressure, etc.) (www.bluetooth.com):

- Bluetooth members: 0xFxxx
- GATT characteristic and object type: 0x2xxx
- GATT declarations: 0x28xx and 0x29xx
- GATT service: 0x18xx
- GATT unit: 0x27xx
- protocol identifier: 0x00xx

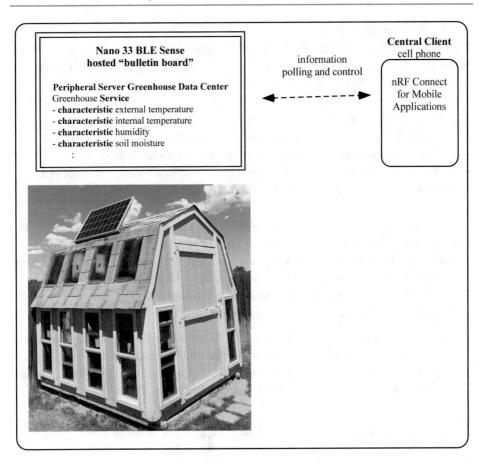

Fig. 2.10 Bluetooth BLE equipped greenhouse

- SDO GATT service: 0XFFFx
- service classes and profiles: 0x10xx and 0x11xx

For BLE services and characteristics without a 16 bit pre–assigned UUID, a unique 128 bit UUID code is used. A Bluetooth unique UUID may be obtained using a number of online UUID generators.

In the greenhouse example, a cell phone is configured as a BLE central or client. Through the BLE wireless radio interconnect, the cell phone can read and interact with the greenhouse data and features.

2.5.3.1 ArduinoBLE Library

The ArduinoBLE Library provides for a wide variety of BLE configurations. The library is downloaded from within the Arduino IDE using the Library Manager. The library is organized into different classes including the (www.arduino.cc):

- BLE Class used to enable the BLE module,
- BLE Device Class to get information about connected devices,
- BLE Service Class to enable services and interaction with services,
- BLE Characteristic Class to enable characteristics and interaction with them, and
- BLE Descriptor Class to describe characteristics.

To get acquainted with the library we continue with a series of examples. The first two examples are adapted from the Arduino BLE Library. In the third example, we configure an Arduino Nano 33 BLE Sense as the server to collect and post greenhouse data. A cell phone is configured as a client to poll and interact with the greenhouse data. The cell phone is equipped with a BLE compatible app to interact with the Nano 33. This example is provided in the Application section at the end of the chapter.

Example: In this first example "LED," from the Arduino BLE Library, a cell phone serves as a central client to control an LED onboard the Nano 33 BLE Sense configured as a server.

To get better acquainted with the sketch, we study the Bluetooth configuration related code steps. In Fig. 2.11 we detail these steps in a UML activity diagram.

```
//*************************************************************
//LED:  This example creates a BLE peripheral with service
//that contains a characteristic to control an LED.
//
//A generic BLE central phone app, like LightBlue or
//nRF Connect is used to interact with the Nano 33
//hosted BLE services and characteristics created in this
//sketch.
//
//This example code is in the public domain.
//*************************************************************

#include <ArduinoBLE.h>
                                  //Declare BLE LED Service
                                  //Link to 128 bit UUID
BLEService ledService("19B10000-E8F2-537E-4F6C-D104768A1214");

//BLE LED Switch Characteristic - custom 128-bit UUID, read and
//writable by central client device (cell phone)
BLEByteCharacteristic switchCharacteristic
   ("19B10001-E8F2-537E-4F6C-D104768A1214", BLERead | BLEWrite);

const int ledPin = LED_BUILTIN;      //Use builtin LED
```

```
void setup()
{
Serial.begin(9600);                    //status to serial monitor
while (!Serial);

pinMode(ledPin, OUTPUT);               //set LED pin to output mode

if(!BLE.begin())                       //BLE initialization
   {
   Serial.println("starting BLE failed!");
   while (1);
   }

   //set advertised local name and service UUID:
   BLE.setLocalName("LED");
   BLE.setAdvertisedService(ledService);

   //add the characteristic to the service
   ledService.addCharacteristic(switchCharacteristic);

   //add service
   BLE.addService(ledService);

   //set the initial value for the characeristic:
   switchCharacteristic.writeValue(0);

   //start advertising
   BLE.advertise();

   Serial.println("BLE LED Peripheral");
}

void loop()
{
//listen for BLE clients (central) to connect:
BLEDevice central = BLE.central();

//if a client (central) is connected to peripheral:
if(central)
   {
   Serial.print("Connected to client: ");

   //print the client's MAC address:
   Serial.println(central.address());

   //while the client (central) is still connected to
   //the Nano 33 based server (peripheral):
   while (central.connected())
      {
      //if the remote client device wrote to the
      //Nano 33 server characteristic, use the
      //value to control the LED:
        if(switchCharacteristic.written())
```

```
      {
    if(switchCharacteristic.value())
      {
      Serial.println("LED on");       //any value other than 0
      digitalWrite(ledPin, HIGH);     //will turn the LED on
      }
    else
      {
      Serial.println(F("LED off")); //a 0 value
      digitalWrite(ledPin, LOW);      //will turn the LED off
      }
    }
  }//end while

  //when the central disconnects, print it out:
  Serial.print(F("Disconnected from central: "));
  Serial.println(central.address());
  }//end if(central)
}

//*************************************************************
```

The sketch may be compiled and uploaded to the Nano 33 BLE. Once uploaded, the sketch may be tested:

- Open the Serial Monitor in the Arduino IDE to monitor sketch status.
- Using a cell phone as a client, open "nRF Connect" to establish Bluetooth BLE connection with the Nano 33 based server.[3]
- Find "LED" in the nRF scanner list.
- Tap "Connect" to connect the client (cell phone) to the server (Nano 33 BLE).
- By selecting "Client" and the up arrow, values may be sent from the client to the server to control the LED.
- Select "Write Value" and "Unsigned."
- Sending a non–zero turns the LED on while sending zero turns the LED off.

Example: In this example, "battery monitor," adapted from the Arduino BLE Library, the Nano 33 BLE Sense is configured as a server. The Nano 33 BLE Sense monitors the analog signal on A0 and posts this characteristic to the server based bulletin board. A cell phone based client equipped with BLE compatible app is used to poll the posted data.

To simulate a battery, a 100 kOhm potentiometer is connected to pin A0. The potentiometer is connected between 3.3 VDC and ground. The potentiometer wiper arm (center terminal) is connected to the A0 pin.

The client/server connection is tested using techniques similar to those provided in the previous example.

---

[3] BLE applications such as nRF connect or LightBlue are available from your cell phone app store.

Fig. 2.11 Bluetooth BLE
configuration

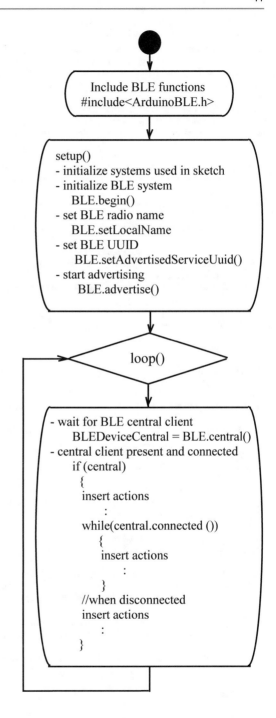

```
//***********************************************************
//Battery Monitor - This example creates a BLE server
//(peripheral) with the standard battery service and level
//characteristic. The A0 pin is used to monitor the battery
//level.
//
//A generic BLE central phone app, like LightBlue or
//nRF Connect is used to interact with the Nano 33
//hosted BLE services and characteristics created in this
//sketch.
//
//This example code is in the public domain.
//***********************************************************

#include <ArduinoBLE.h>

BLEService batteryService("180F");          //BLE Battery Service

//BLE Battery Level Characteristic
//   - standard 16-bit characteristic UUID
//   - remote clients get notifications if characteristic changes
BLEUnsignedCharCharacteristic batteryLevelChar("2A19",
                                       BLERead | BLENotify);
int oldBatteryLevel = 0;                //last battery level reading from A0
long previousMillis = 0;                  //last time battery level checked (ms)

void setup()
{
Serial.begin(9600);                        //initialize serial communication
while (!Serial);

                                           //initialize built-in LED pin
pinMode(LED_BUILTIN,OUTPUT);               //indicates when central is connected

if(!BLE.begin())                           //initialize Bluetooth BLE device
  {
  Serial.println("starting BLE failed!");
  while (1);
  }

//Set a local name for the BLE device. This name appears
//in advertising packets. Name used by remote devices to
//identify this BLE device.
BLE.setLocalName("BatteryMonitor");
BLE.setAdvertisedService(batteryService);    //add the service UUID
                                             //add the battery level characteristic
batteryService.addCharacteristic(batteryLevelChar);
BLE.addService(batteryService);              //add battery service
batteryLevelChar.writeValue(oldBatteryLevel); //set initial value

//Start advertising BLE.  Continuously transmits BLE advertising
//packets. Advertising will be visible to remote BLE central devices.
BLE.advertise();
Serial.println("Bluetooth device active, waiting for connections...");
}

void loop()
{
BLEDevice central = BLE.central();            //wait for a BLE central
                                              //if a central client
```

```
                                        //is connected to peripheral
if(central)
{
Serial.print("Connected to central: ");
Serial.println(central.address());      //print the central's BT address
digitalWrite(LED_BUILTIN, HIGH);        //LED on when client connected
                                        //while client connected
                                        //check battery level every 200ms
while(central.connected())
  {
  long currentMillis = millis();
                                        //if 200ms have passed,
                                        //check the battery level
  if(currentMillis - previousMillis >= 200)
    {
    previousMillis = currentMillis;
    updateBatteryLevel();
    }
  }

  //when the central client disconnects, turn off the LED
  digitalWrite(LED_BUILTIN, LOW);
  Serial.print("Disconnected from central: ");
  Serial.println(central.address());
  }
}

//*************************************************************
//void updateBatteryLevel() - Read the current voltage level
//on the A0 analog input pin. This is used here to simulate
//the battery level.
//*************************************************************

void updateBatteryLevel()
{
int battery = analogRead(A0);
int batteryLevel = map(battery, 0, 1023, 0, 100);

if(batteryLevel != oldBatteryLevel)
  {                                     //if the battery level has changed
  Serial.print("Battery Level %: ");    // print it
  Serial.println(batteryLevel);
  batteryLevelChar.writeValue(batteryLevel);  //update battery level
                                        //characteristic
  oldBatteryLevel = batteryLevel;       //save level for comparison
  }
}

//*************************************************************
```

2.6   Nano 33 BLE Sense Peripherals

As discussed earlier in the chapter, the Arduino Nano 33 BLE Sense SoC board is equipped with a rich complement of sensors. In this section we take a closer look at the following sensors (ABX00031):

- Nine axis inertial measurement unit (IMU) (LSM9DS1),
- Barometer and temperature sensor (LPS22HB),
- Relative humidity sensor (HTS221),
- Digital proximity, ambient light, RGB, and gesture sensor (APDS–9960),
- Digital microphone (MP34DT05).

2.6.1 Nine Axis IMU (LSM9DS1)

It is an important feature to locally sense orientation in the X, Y, Z coordinate system An inertial measurement unit (IMU) provides this capability. An IMU consists of an accelerometer, gyroscope, and magnetometer to measure acceleration, rotation, and the Earth's magnetic field strength in the X, Y, Z coordinate system as shown in Fig. 2.12.

The Nano 33 BLE sense is equipped with the LSM9DS1 iNemo inertial module. The module consists of a 3D accelerometer, 3D gyroscope, and a 3D magnetometer. The module has the following specifications (LSM9DS1):

- $\pm$ 2, 4, 8, 16 g linear acceleration,
- $\pm$ 4, 8, 12, 16 Gauss magnetic full scale, and
- $\pm$ 245, 500, 2000 degree per second (dps) angular rate full scale.

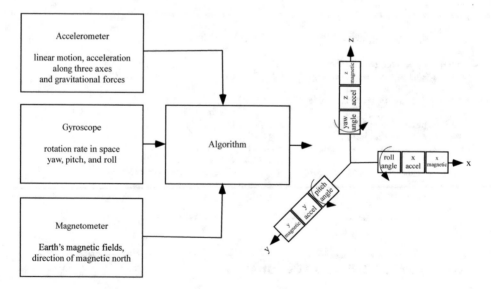

Fig. 2.12 Inertial measurement unit

Example: The "Arduino_LSM9DS1" library created by Riccardo Rizzo provides examples to exercise the onboard accelerometer, gyroscope, and magnetometer. Provided here is the "Simple_Accelerometer."

```
//*********************************************************
//Arduino LSM9DS1 - Simple Accelerometer
//This example reads the acceleration values from
//the LSM9DS1 sensor and continuously prints them
//to the Serial Monitor or Serial Plotter.
//
//The circuit: Arduino Nano 33 BLE Sense
//
//created 10 Jul 2019 by Riccardo Rizzo
//
//This example code is in the public domain.
//*********************************************************

#include <Arduino_LSM9DS1.h>

void setup()
{
Serial.begin(9600);
while (!Serial);
Serial.println("Started");

if(!IMU.begin())
   {
   Serial.println("Failed to initialize IMU!");
   while (1);
   }

Serial.print("Accelerometer sample rate = ");
Serial.print(IMU.accelerationSampleRate());
Serial.println(" Hz");
Serial.println();
Serial.println("Acceleration in G's");
Serial.println("X\tY\tZ");
}

void loop()
{
float x, y, z;

if(IMU.accelerationAvailable())
   {
   IMU.readAcceleration(x, y, z);

   Serial.print(x);
   Serial.print('\t');
   Serial.print(y);
```

```
    Serial.print('\t');
    Serial.println(z);
    }
}


//***************************************************
```

2.6.2 Barometer and Temperature Sensor (LPS22HB)

The LPS22HB sensor module provides barometric pressure and temperature. Pressure (in Pascals) is the measurement of force (Newtons) per area ($meter^2$). Barometric pressure is the pressure on Earth caused by the weight of air above as compared to a vacuum. Intuitively, barometric pressure decreases as the altitude where the measurement is taken increases. Also, barometric pressure varies depending on weather conditions. For example, the barometric pressure decreases in the presence of rainy weather [6].

The LPS22HB provides pressure readings from 260 to 1260 hPa (hecto Pascals). The Standard Atmosphere at sea level is given as 1013 hPa. This is equivalent to 101.325 kPa and 1013.25 mbar and 14.696 PSI (LPS22HB).

Example: The example "ReadPressure" within the "Arduino_LPS22HB" Library provides a reading of pressure and temperature. Interestingly, when I run this program in my home lab, I receive a reading of 77.62 kPa. Recall, the Standard Atmosphere at sea level is given as 101.325 kPa. When the 77.62 kPa reading is provided to an "Air Pressure at Altitude Calculator," my altitude is given as 7,194.33 feet (www.mide.com). The University of Wyoming football and basketball teams regularly reminds opponents they are competing at 7,220 feet!

```
//***************************************************
//LPS22HB - Read Pressure
//This example reads data from the on-board LPS22HB
//sensor of the Nano 33 BLE Sense and prints the
//temperature and pressure sensor value to the Serial
//Monitor once a second.
//
//The circuit: Arduino Nano 33 BLE Sense
//
//This example code is in the public domain.
//***************************************************

#include <Arduino_LPS22HB.h>

void setup()
{
Serial.begin(9600);
while (!Serial);

if(!BARO.begin())
```

```
  {
  Serial.println("Failed to initialize pressure sensor!");
  while (1);
  }
}

void loop()
{
//read the sensor value
float pressure = BARO.readPressure();

//print the sensor value
Serial.print("Pressure = ");
Serial.print(pressure);
Serial.println(" kPa");

float temperature = BARO.readTemperature();

//print the sensor value
Serial.print("Temperature = ");
Serial.print(temperature);
Serial.println(" C");

  // print an empty line
  Serial.println();

  // wait 1 second to print again
  delay(1000);
}

//*****************************************************
```

2.6.3 Relative Humidity and Temperature Sensor (HTS221)

The HTS221 sensor provides for measuring relative humidity (rH). Relative humidity provides a measure of water vapor content within air. Absolute humidity is the mass of water vapor per volume of air measured in $grams/meter^3$. Relative humidity references absolute humidity to the amount of water vapor the volume of air could contain at a given temperature. The relative humidity is reported as a percentage between 0 and 100 percent [6].

I have lived in many places that exhibited a wide range of humidity. For example, I attended junior high school on Guam. Guam is a tropical island in the Pacific Ocean where relative humidity regularly exceeds 80 percent. I now live in Wyoming where the relative humidity during the summer averages 25 percent. Figure 2.13 illustrates the range of relative humidity.

The HTS221 sensor provides rH from 0 to 100 percent. It provides an accuracy of ± 3.5% in the range of 20 to 80% rH (HTS221).

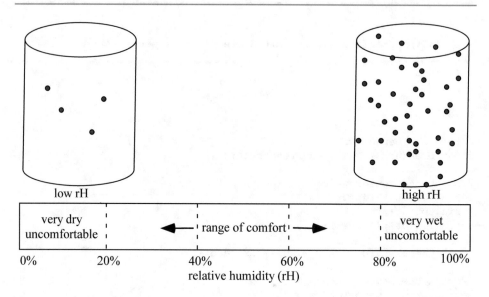

Fig. 2.13 Relative humidity (rH) [6]

Example: The example "ReadSensors" within the "Arduino_HTS221" Library provides a reading of relative humidity and temperature.

```
//*****************************************************
//HTS221 - Read Sensors
//This example reads data from the on-board HTS221
//sensor of the Nano 33 BLE Sense and prints the
//temperature and humidity sensor values to the Serial
//Monitor once a second.
//
//The circuit: Arduino Nano 33 BLE Sense
//
//This example code is in the public domain.
//*****************************************************

#include <Arduino_HTS221.h>

void setup()
{
Serial.begin(9600);
while (!Serial);

if(!HTS.begin())
  {
  Serial.println("Failed to initialize humidity temperature sensor!");
  while (1);
  }
```

```
}

void loop()
{
//read all the sensor values
float temperature = HTS.readTemperature();
float humidity    = HTS.readHumidity();

//print each of the sensor values
Serial.print("Temperature = ");
Serial.print(temperature);
Serial.println(" Â°C");

Serial.print("Humidity    = ");
Serial.print(humidity);
Serial.println(" %");

//print an empty line
Serial.println();

//wait 1 second to print again
delay(1000);
}

//********************************************************
```

2.6.4   Digital Proximity, Ambient Light, RGB, and Gesture Sensor (APDS–9960)

The Arduino Nano 33 BLE Sense is equipped with the APDS–9960 sensor. The sensor is equipped with gesture detection, color component, and proximity detection features. We discuss each in turn with an accompanying example sketch.

2.6.4.1 Gesture Detection

The gesture detection feature uses four directional photodiodes as shown in Fig. 2.14a. Gesture direction is defined as shown. The LED shown below the photodiodes provides the source illumination. When a gesture is made, for example a left to right sweep in front of the photodiodes, a right gesture is detected.

Arduino provides the ADPS–9960 Library which includes sketches to test the different features.

Example: In this example we have modified the Gesture Sensor sketch provided in the APDS–9960 Library. When a gesture is detected a corresponding LED is illuminated. The configuration of the display LEDs is shown in Fig. 2.14b.

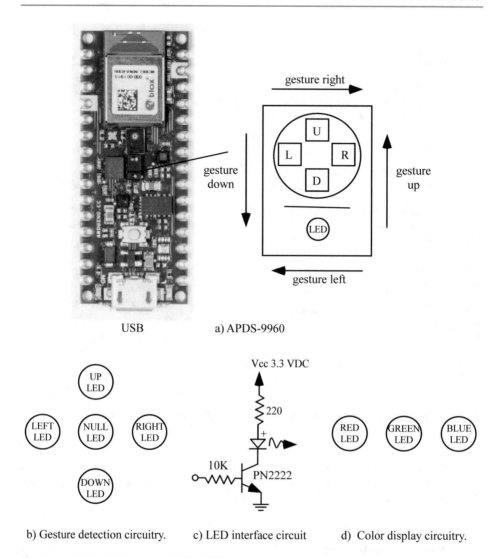

b) Gesture detection circuitry.   c) LED interface circuit   d) Color display circuitry.

Fig. 2.14 APDS–9960 sensor (APDS–9960)

```
//*******************************************************************
//APDS-9960 - Gesture Sensor - This example reads gesture data from
//the onboard APDS-9960 sensor of the Nano 33 BLE Sense and prints
//any detected gestures to the Serial Monitor and illuminates a
//corresponding LED to indicate the gesture direction.
//
//Gesture directions are defined as:
//- UP:    from USB connector towards antenna
//- DOWN:  from antenna towards USB connector
//- LEFT:  from analog pins side towards digital pins side
//- RIGHT: from digital pins side towards analog pins side
//
```

```
//Five LEDs are used to indicate gesture direction:
//- UP LED:    Digital pin D2 - physical pin 20
//- DOWN LED:  Digital pin D3 - physical pin 21
//- LEFT LED:  Digital pin D4 - physical pin 22
//- Right LED: Digital pin D5 - physical pin 23
//- NULL LED:  Digital pin D6 - physical pin 24
//
//The example code "Gesture Sensor" from the Arduino APDS-9960
//Library was adapted to include the five LED indicators.  The code
//is in the public domain.
//****************************************************************


#include <Arduino_APDS9960.h>

#define UP_LED    2        //physical pin
#define DOWN_LED  3        //physical pin
#define LEFT_LED  4        //physical pin
#define RIGHT_LED 5        //physical pin
#define NULL_LED  6        //physical pin

void setup()
{
pinMode(UP_LED,    OUTPUT);
pinMode(DOWN_LED, OUTPUT);
pinMode(LEFT_LED, OUTPUT);
pinMode(RIGHT_LED,OUTPUT);
pinMode(NULL_LED, OUTPUT);

Serial.begin(9600);
while (!Serial);

if(!APDS.begin())
  {
  Serial.println("Error initializing APDS-9960 sensor!");
  }

//Note: for setGestureSensitivity(..) a value between 1 and 100
//is required.  Higher values make the gesture recognition more
//sensitive but less accurate (i.e. a wrong gesture may be
//detected). Lower values makes the gesture recognition more accurate
//but less sensitive (i.e. some gestures may be missed).  A default
//of 80 is used.  To change this value, uncomment the next line and insert a desired threshold value.
//APDS.setGestureSensitivity(80);

Serial.println("Detecting gestures ...");
}

void loop()
{
//LED setting to default values
digitalWrite(UP_LED,    LOW);
digitalWrite(DOWN_LED, LOW);
digitalWrite(LEFT_LED, LOW);
digitalWrite(RIGHT_LED,LOW);
digitalWrite(NULL_LED, HIGH);

if(APDS.gestureAvailable())        //a gesture was detected,
  {                                //print to Serial Monitor
                                   //illuminate LEDs
    int gesture = APDS.readGesture();

    switch(gesture)
      {
        case GESTURE_UP:
        Serial.println("Detected UP gesture");
        digitalWrite(UP_LED,    HIGH);
```

```
      digitalWrite(NULL_LED, LOW);
      break;

      case GESTURE_DOWN:
      digitalWrite(DOWN_LED, HIGH);
      digitalWrite(NULL_LED, LOW);
      Serial.println("Detected DOWN gesture");
      break;

      case GESTURE_LEFT:
      Serial.println("Detected LEFT gesture");
      digitalWrite(LEFT_LED, HIGH);
      digitalWrite(NULL_LED, LOW);
      break;

      case GESTURE_RIGHT:
      Serial.println("Detected RIGHT gesture");
      digitalWrite(RIGHT_LED, HIGH);
      digitalWrite(NULL_LED, LOW);
      break;

      default:      //ignore
      break;


   }
   delay(1000); //1 second delay
   }
}

//***********************************************************************
```

2.6.4.2 Color Sensor

The APDS–9960 is also equipped with a color sensor that reports on the chromatic content (red, green, blue) of detected light. The different color content is sensed using three different photodiodes detecting red (centered at approximately 610 nm), green (centered at approximately 540 nm), and blue (centered at approximately 450 nm).

The following sketch "Color Sensor," provided in the Arduino ADPS–9960 Library, is adapted to illuminate three separate LEDs at the intensity of the detected chromatic content. The R, G, B LED circuit is shown in Fig. 2.14c. The sketch is tested by placing different color filters in front of a light source (e.g. Neewer 58 mm full color lens filter set).

```
//***********************************************************************
//APDS-9960 - Color Sensor1 - This example reads color data from
//the onboard APDS-9960 sensor of the Nano 33 BLE Sense and prints
//the color RGB (red, green, blue) values to the Serial Monitor once
//a second.  Also, external R, G, and B LEDs are illuminated with the
//corresponding intensity of the R,G, B component reading.
//
//Three LEDs are used to indicate R,G,B component presence.
//- RED LED:    Digital PWM pin D3
//- BLUE LED:   Digital PWM pin D5
//- GREEN LED:  Digital PWM pin D6
//
//The example code "Color Sensor" from the Arduino APDS-9960 Library
```

```
//was adapted to include the three LED indicators.  The code is in
//the public domain.
//*******************************************************************

#include <Arduino_APDS9960.h>

#define RED_LED    3        //physical pin 21
#define BLUE_LED   5        //physical pin 23
#define GREEN_LED  6        //physical pin 24

void setup()
{
pinMode(RED_LED,    OUTPUT);
pinMode(BLUE_LED,   OUTPUT);
pinMode(GREEN_LED,  OUTPUT);

Serial.begin(9600);
while (!Serial);

if(!APDS.begin())
  {
  Serial.println("Error initializing APDS-9960 sensor.");
  }
}

void loop()
{
//LED setting to default values
digitalWrite(RED_LED,    LOW);
digitalWrite(BLUE_LED,   LOW);
digitalWrite(GREEN_LED,  LOW);

//check if a color reading is available
while (! APDS.colorAvailable())
  {
  delay(5);
  }
  int r, g, b;

//read the color
APDS.readColor(r, g, b);

//print the values
Serial.print("r = ");
Serial.println(r);
Serial.print("g = ");
Serial.println(g);
Serial.print("b = ");
Serial.println(b);
Serial.println();

//illuminate the LEDs to intensity of component value via PWM
analogWrite(RED_LED,    r);
```

```
analogWrite(GREEN_LED, g);
analogWrite(BLUE_LED,  b);


delay(100);                           //wait a bit before reading again
}


//******************************************************************
```

2.6.4.3 Proximity Sensor

The APDS–9960 is also equipped with a proximity sensor. The sensor uses an onboard infrared diode as an illumination source. The light source is reflected off an object. The reflected light is collected by the same four photodiodes used for gesture detection. A numerical value corresponding to the range of the object is reported by the ADPS–9960. The numerical value may be correlated with a specific range for a given application.

The following sketch "Proximity Sensor," provided in the Arduino ADPS–9960 Library, is adapted to report an object's range and also illuminate an LED at an intensity consistent with range (i.e. closer object, brighter LED).

```
//**********************************************************************
//APDS-9960 - Proximity Sensor - This example reads proximity data
//from the onboard APDS-9960 sensor of the Nano 33 BLE Sense and
//prints the proximity value to the Serial Monitor every 100 ms.
//Sensed values range from 0 (close) to 255 (far).  The sensed value
//is used to set the intensity of an LED using PWM.  The LED is
//connected to D3.
//
//The example code "Proximity Sensor" from the Arduino APDS-9960
//Library was adapted to include the LED indicator.  The code is in
//the public domain.
//**********************************************************************

#include <Arduino_APDS9960.h>

#define  range_LED   3   //physical pin 21

void setup()
{
pinMode(range_LED,   OUTPUT);
Serial.begin(9600);
while (!Serial);

if(!APDS.begin())
  {
  Serial.println("Error initializing APDS-9960 sensor!");
  }
}

void loop()
```

```
{
//LED setting to default value
digitalWrite(range_LED,    LOW);

//check if a proximity reading is available
if(APDS.proximityAvailable())
  {
  // read the proximity
  // - 0   => close
  // - 255 => far
  // - -1  => error
  int proximity = APDS.readProximity();

  // print value to the Serial Monitor
  Serial.println(proximity);

  //illuminate LED to appropriate intensity
  //closer proximity, brighter LED
  proximity = abs(proximity - 255);

  analogWrite(range_LED, proximity);
  }

  // wait a bit before reading again
  delay(100);
}

//*********************************************************************
```

2.6.5   Digital Microphone (MP34DT05)

The Nano 33 BLE Sense is equipped with an omnidirectional microphone. This means in can accept sound from multiple directions. It uses a pulse density modulation (PDM) technique where the sound is encoded into a serial stream of digital pulses. The digital pulses may be averaged to render an analog audio output. The microphone's signal to noise ratio (SNR) is 64 dB with a sensitivity of –26 dBFS (MP34DT05).

Example: The "PDMSerialPlotter" sketch captures samples from the MP34DT05 digital microphone and displays the samples in the Serial Monitor. The samples can also be displayed as a time sequence by choosing Tools –> Serial Plotter within the Arduino IDE.

```
//*****************************************************
//PDMSerialPlotter: reads audio data from the on-board
//PDM microphones, and prints out the samples to the
//Serial Monitor. The Serial Plotter built into the
//Arduino IDE can be used to plot the audio data
//(Tools -> Serial Plotter)
//
//Circuit: Arduino Nano 33 BLE board
```

```
//
//This example code is in the public domain.
//****************************************************

#include <PDM.h>

//default number of output channels
static const char channels = 1;

//default PCM output frequency
static const int frequency = 16000;

//Buffer to read samples, each sample is 16-bits
short sampleBuffer[512];

//Number of audio samples read
volatile int samplesRead;

void setup()
{
Serial.begin(9600);
while (!Serial);

//Configure the data receive callback
PDM.onReceive(onPDMdata);

//Optionally set the gain.
//Defaults to 20 on the BLE Sense
//PDM.setGain(30);

//Initialize PDM with:
// - one channel (mono mode)
// - a 16 kHz sample rate for the Arduino Nano 33 BLE Sense
if(!PDM.begin(channels, frequency))
  {
  Serial.println("Failed to start PDM!");
  while (1);
  }
}

void loop()
{
//Wait for samples to be read
if(samplesRead)
  {
  //Print samples to the serial monitor or plotter
  for(int i = 0; i < samplesRead; i++)
    {
    if(channels == 2)
      {
      Serial.print("L:");
      Serial.print(sampleBuffer[i]);
      Serial.print(" R:");
```

```
        i++;
      }
    Serial.println(sampleBuffer[i]);
    }

  //Clear the read count
  samplesRead = 0;
  }
}


//*******************************************************
// void onPDMdata(): callback function to process the
//data from the PDM microphone.
//Note: This callback is executed as part of an ISR.
//Therefore using 'Serial' to print messages inside
//this function isn't supported.
//*******************************************************

void onPDMdata()
{
//Query the number of available bytes
int bytesAvailable = PDM.available();

//Read into the sample buffer
PDM.read(sampleBuffer, bytesAvailable);

//16-bit, 2 bytes per sample
samplesRead = bytesAvailable / 2;
}


//*******************************************************
```

2.7   Application: Bluetooth BLE Greenhouse Monitor

In developing software for a system, it is not always possible or desirable to have close
access to the system. For example, in the development of control software for a greenhouse,
the greenhouse is not always readily available. Also, different weather conditions are not
conveniently available to test the control algorithm under a variety of conditions. In these
situations a simulator may be used to substitute for the system. The simulator provides the
necessary inputs and signals in place of the system so that software may be developed.

In the following example, we develop a Bluetooth BLE application to gather greenhouse
data and make it available for viewing on a client cell phone. A small hardware simulator
consisting of several potentiometers and LED indicators are used as a greenhouse substitute
during software development.

The hardware simulator is shown in Fig. 2.15. A series of seven 100 kOhm potentiometers
provide simulated weather data. One side of each potentiometer is connected to Vcc and the
other side to ground. The wiper for a given potentiometer is connected to a jumper wire for

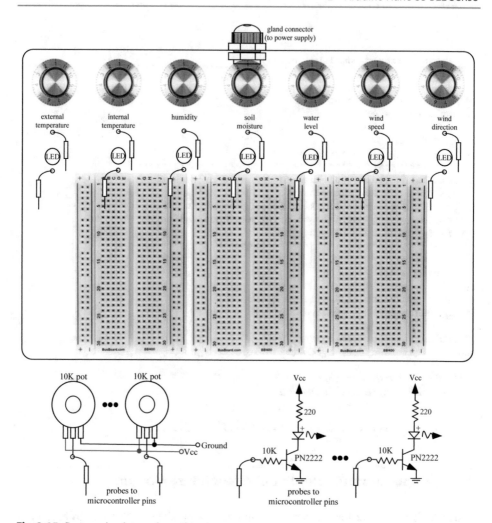

Fig. 2.15 System simulator schematic

easy connection to a microcontroller pin via the prototype board. A series of seven LEDs are used to simulate motors, pumps, etc. Each LED is equipped with an interface circuit as shown in Fig. 2.15. The input to each interface circuit is provided a jumper wire. The completed simulator is shown in Fig. 2.16.

In "Arduino III: Internet of Things," a design is provided for a greenhouse and accompanying weather station. The greenhouse control system uses a 3.3 VDC Arduino MKR 1000 microcontroller. We use the simulator in place of the greenhouse. In this example we use the Arduino Nano 33 BLE Sense as a server for greenhouse parameters and make them available to BLE peripheral clients. A cell phone serves as a client. Through a BLE app (e.g. nRF Connect, LightBlue), the cellphone is used to read greenhouse parameters and to control a

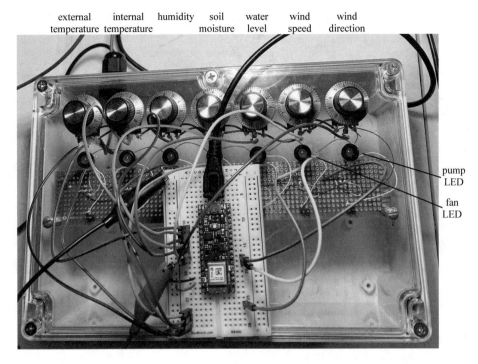

external   internal   humidity   soil   water   wind   wind
temperature temperature   moisture   level   speed   direction

pump
LED

fan
LED

Fig. 2.16 System simulator

simulated vent fan or water pump. Lessons learned from the BLE examples provided earlier
in the chapter are used as building blocks for this more complex control algorithm.

```
//********************************************************************
//GreenHouse Monitor:  This example provides a BLE peripheral to
//monitor parameters and control features including:
//
//Greenhouse (GH) Service (0x1800)
//Greenhouse Characteristics
//- External temperature (0x2A1C) - A0 (pin 4)
//- Internal temperature (0x2A6E) - A1 (pin 5)
//- Humidity (0x2A6F) - A2 (pin 6)
//- Soil moisture (0x2ACA) - A3 (pin 7)
//- Water level in rain barrel  (0x2A78) - A4 (pin 8)
//- Wind speed (0x2A70) - A5 (pin 9)
//- Wind direction (0x2A71) - A6 (pin 10)
//- Battery level (0x2A19) - A7 (pin 11)
//- Vent fan ("19B10001-E8F2-537E-4F6C-D104768A121F") - D2 (pin 20)
//- Water pump ("19B10001-E8F2-537E-4F6C-D104768A121A") - D3 (pin 21)
//Standard BLE characteristic and service codes are used where
//available.
//
//You can use a generic BLE central app (e.g. LightBlue, nRF Connect),
//to interact with the services and characteristics created in this
//sketch.
//
//This example code is in the public domain.
```

```
//*********************************************************************

#include <ArduinoBLE.h>
                                        //Declare BLE service
                                        //BLE Greenhouse service
BLEService greenhouseService("19B10001-E8F2-537E-4F6C-D104768A1214"); //BLE GH Characteristic
//External temperature (0x2A1C) - A0 (pin 4)
BLEUnsignedCharCharacteristic exttempChar("2A1C", BLERead | BLENotify);
int oldexttemp = 0;        //last ext temp reading from analog input A0

//Internal temperature (0x2A6E) - A1 (pin 5)
BLEUnsignedCharCharacteristic inttempChar("2A6E", BLERead | BLENotify);
int oldinttemp = 0;        //last int temp reading from analog input A1

//Humidity (0x2A6F) - A2 (pin 6)
BLEUnsignedCharCharacteristic humidityChar("2A6F", BLERead | BLENotify);
int oldhumidity = 0;       //last humidity reading from analog input A2

//Soil moisture (0x2ACA) - A3 (pin 7)
BLEUnsignedCharCharacteristic soilmoistureChar("2ACA",BLERead |
                                       BLENotify);
int oldsoilmoisture = 0;   //last soil moisture reading analog input A3

//Water level in rain barrel  (0x2A78) - A4 (pin 8)
BLEUnsignedCharCharacteristic rainlevelChar("2A78", BLERead | BLENotify);
int oldrainlevel = 0;      //last rain level reading from analog input A4

//Wind speed (0x2A70) - A5 (pin 9)
BLEUnsignedCharCharacteristic windspeedChar("2A70", BLERead | BLENotify);
int oldwindspeed = 0;      //last windspeed reading from analog input A5

//Wind direction (0x2A71) - A6 (pin 10)
BLEUnsignedCharCharacteristic winddirectionChar("2A71",BLERead |
                                       BLENotify);
int oldwinddirection = 0; //last wind direction reading analog input A6

//Battery level (0x2A19) - A7 (pin 11)
BLEUnsignedCharCharacteristic batteryLevelChar("2A19", BLERead |
                                       BLENotify);
int oldBatteryLevel = 0;  //last battery level reading analog input A7

long previousMillis = 0;  //last time the parameters were checked, ms

//Vent fan - client can remotely operate vent fan
BLEByteCharacteristic switchCharacteristic1
            ("19B10001-E8F2-537E-4F6C-D104768A121F", BLERead | BLEWrite);
const int fan_ledPin = 2; //pin to use for the fan LED

//Water pump - client can remotely operate water pump for mister
BLEByteCharacteristic switchCharacteristic2
            ("19B10001-E8F2-537E-4F6C-D104768A121A", BLERead | BLEWrite);
const int pump_ledPin = 3; //pin to use for the water pump LED

void setup()
{
Serial.begin(9600);                    //initialize serial communication
while (!Serial);
pinMode(fan_ledPin, OUTPUT);           //initialize LED - fan
                                       //indicates when central connected

if(!BLE.begin())                       //begin initialization
  {
  Serial.println("starting BLE failed!");
  while (1);
  }
```

```
//Set local name for BLE device. Name will appear in advertising packets.
//Used by remote devices to identify BLE device.
BLE.setLocalName("Greenhouse Monitor");
BLE.setAdvertisedService(greenhouseService);                //add the service UUID
greenhouseService.addCharacteristic(exttempChar);           //add exttemp char
greenhouseService.addCharacteristic(inttempChar);           //add inttemp char
greenhouseService.addCharacteristic(humidityChar);          //add humidity char
greenhouseService.addCharacteristic(soilmoistureChar);      //add soil moisture
greenhouseService.addCharacteristic(rainlevelChar);         //add rain level char
greenhouseService.addCharacteristic(windspeedChar);         //add windspeed char
greenhouseService.addCharacteristic(winddirectionChar);     //wind dir char
greenhouseService.addCharacteristic(batteryLevelChar);      //batt level char
greenhouseService.addCharacteristic(switchCharacteristic1);//fan char
greenhouseService.addCharacteristic(switchCharacteristic2);//pump char

BLE.addService(greenhouseService);                          //add the GH service
exttempChar.writeValue(oldexttemp);                         //set initial value  char
inttempChar.writeValue(oldinttemp);
humidityChar.writeValue(oldhumidity);
soilmoistureChar.writeValue(oldsoilmoisture);
rainlevelChar.writeValue(oldrainlevel);
windspeedChar.writeValue(oldwindspeed);
winddirectionChar.writeValue(oldwinddirection);
batteryLevelChar.writeValue(oldBatteryLevel);

switchCharacteristic1.writeValue(0);            //vent fan off
BLEDescriptor vent_fanDescriptor("19B10001-E8F2-537E-4F6C-D104768A121F", "Vent Fan");
switchCharacteristic1.addDescriptor(vent_fanDescriptor);

switchCharacteristic2.writeValue(0);            //water pump off
BLEDescriptor water_pumpDescriptor
                ("19B10001-E8F2-537E-4F6C-D104768A121A", "Water Pump");
switchCharacteristic2.addDescriptor(water_pumpDescriptor);


BLE.advertise();                                //start advertising
Serial.println("Bluetooth device active, waiting for connections...");
}

void loop()
{
BLEDevice central = BLE.central();              //wait for a BLE central

if(central)                                     //if central client connected...
  {
  Serial.print("Connected to central: ");
  Serial.println(central.address());            //print clients's BT address

                                                //while client is connected
  while(central.connected())
    {
    long currentMillis = millis();
    if(currentMillis - previousMillis >= 200) //check param every 200 ms
      {
      previousMillis = currentMillis;
      update_sensor_values();
      }

    //if the remote device wrote to the characteristic,
    //use the value to control the fan
    if(switchCharacteristic1.written())
      {
      if(switchCharacteristic1.value())
        {                                       //any value other than 0
```

```
            Serial.println("Vent Fan - on");
            digitalWrite(fan_ledPin, HIGH);      //will turn the LED (fan) on
            }
        else
            {                                     //a 0 value
            Serial.println(F("Vent Fan - off"));
            digitalWrite(fan_ledPin, LOW);        //will turn the LED (fan) off
            }
        }

    //if the remote device wrote to the characteristic,
    //use the value to control the pump
    if(switchCharacteristic2.written())
        {
        if(switchCharacteristic2.value())
            {                                     //any value other than 0
            Serial.println("Water Pump - on");
            digitalWrite(pump_ledPin, HIGH);      //will turn the LED (pump) on
            }
        else
            {                                     //a 0 value
            Serial.println(F("Water Pump - off"));
            digitalWrite(pump_ledPin, LOW);       //turn the LED (pump) off
            }
        }

    }//end while

    Serial.print("Disconnected from central: ");
    Serial.println(central.address());
  }
}

//*****************************************************************
//void update_sensor_values() - Read the current sensor values
//on the analog input pins.
//*****************************************************************

void update_sensor_values()
{
//External greenhouse temperature
int exttemp = analogRead(A0);
int exttempLevel = map(exttemp, 0, 1023, 0, 100);
if (exttempLevel != oldexttemp)
    {                                            //ext temp level changed
    Serial.print("Ext Temp Level % is now: "); //print it
    Serial.println(exttempLevel);
    exttempChar.writeValue(exttempLevel);        //update ext temp level char
    oldexttemp = exttempLevel;                   //save level for next comparison
    }

//Internal greenhouse temperature
int inttemp = analogRead(A1);
int inttempLevel = map(inttemp, 0, 1023, 0, 100);
if (inttempLevel != oldinttemp)
    {                                            //if int temp level changed
    Serial.print("Int Temp Level % is now: "); //print it
    Serial.println(inttempLevel);
    inttempChar.writeValue(inttempLevel);        //update int temp level char
    oldinttemp = inttempLevel;                   //save level for next comp
    }

//Internal greenhouse humidity
int inthumidity = analogRead(A2);
int inthumidityLevel = map(inthumidity, 0, 1023, 0, 100);
```

```
if (inthumidityLevel != oldhumidity)
   {                                          //if int humidity changed
   Serial.print("Int Humidity % is now: ");   //print it
   Serial.println(inthumidityLevel);
   humidityChar.writeValue(inthumidityLevel); //update int humidity level char
   oldhumidity = inthumidityLevel;            //save level for next comparison
   }

//Internal soil moisture
int intsoilmoisture = analogRead(A3);
int intsoilmoistureLevel = map(intsoilmoisture, 0, 1023, 0, 100);
if (intsoilmoistureLevel != oldsoilmoisture)
   {                                          //if level changed
   Serial.print("Soil Moisture Level % is now: ");    //print it
   Serial.println(intsoilmoistureLevel);
   soilmoistureChar.writeValue(intsoilmoistureLevel); //update char
   oldsoilmoisture = intsoilmoistureLevel;    //save for next comparison
   }

//Rain level in internal capture barrel
int intrainlevel = analogRead(A4);
int intrainlevelLevel = map(intrainlevel, 0, 1023, 0, 100);
if (intrainlevelLevel != oldrainlevel)
   {                                          //if level changed
   Serial.print("Rain Barrel Level % is now: ");//print it
   Serial.println(intrainlevelLevel);
   rainlevelChar.writeValue(intrainlevelLevel); //update rain level char
   oldrainlevel = intrainlevelLevel;          //save for next comp
   }

//Windspeed
int intwindspeed = analogRead(A5);
int intwindspeedLevel = map(intwindspeed, 0, 1023, 0, 100);
if (intwindspeedLevel != oldwindspeed)
   {                                          //if wind speed changed
   Serial.print("Wind Speed % is now: ");     //print it
   Serial.println(intwindspeedLevel);
   windspeedChar.writeValue(intwindspeedLevel); //update wind level char
   oldwindspeed = intwindspeedLevel;          //save for next comp
   }

//Wind direction
int intwinddirection = analogRead(A6);
int intwinddirectionLevel = map(intwinddirection, 0, 1023, 0, 100);
if (intwinddirectionLevel != oldwinddirection)
   {                                          //if direction changed
   Serial.print("Wind Direction % is now: "); //print it
   Serial.println(intwinddirectionLevel);
   winddirectionChar.writeValue(intwinddirectionLevel);//update char
   oldwinddirection = intwinddirectionLevel;  //save for next comp
   }

int battery = analogRead(A7);
int batteryLevel = map(battery, 0, 1023, 0, 100);

if (batteryLevel != oldBatteryLevel)
   {                                          //if level changed
   Serial.print("Battery Level % is now: ");  //print it
   Serial.println(batteryLevel);
   batteryLevelChar.writeValue(batteryLevel); //update level char
   oldBatteryLevel = batteryLevel;            //save for next comp
   }
}

//****************************************************************
```

2.8   Summary

In this chapter we explored the Arduino Nano 33 BLE Sense SoC board and its nRF52840 processor. We began with an overview of the Arduino Nano 33 BLE Sense SoC board features. We then examined the powerful and well–equipped nRF52840 processor and its associated peripheral subsystems. We then investigated the multiple peripherals onboard the Nano 33 BLE Sense SoC board. For selected peripherals we provided a brief theory of operation, feature overview, and examples. We concluded the chapter with an extended example featuring a Bluetooth BLE application–a greenhouse monitoring system.

2.9   Problems

1. What is the operating voltage of the Arduino Nano 33 BLE Sense? What constraints does this operating voltage put on the processor?
2. The onboard RGB LEDs are active low? What does this mean?
3. How many external pins are available on the Arduino Nano 33 BLE Sense? What external pin is used to provide power to the Nano 33? What is the acceptable range of this input voltage?
4. Provide a list of onboard Nano 33 BLE sensors. Provide a brief description of each sensor.
5. Briefly describe the serial communication features onboard the Nano 33 BLE sense. Provide a summary table of features.
6. Briefly describe the difference between EEPROM Flash and SRAM memory. How are these memory elements used in a typical program execution?
7. What is PWM? How is it used to control the speed of a motor? The intensity of an LED?
8. What is the difference between an ADC and DAC system? How are these systems used in a microcontroller application?
9. What is the difference between Bluetooth Classic and BLE? Are they compatible with one another?
10. Briefly describe the role of the client and server in a Bluetooth BLE application?
11. What is a Bluetooth UUID?
12. What is the advantage of using a system simulator in Arduino sketch development?
13. What is the function of an inertial measurement unit?
14. What is barometric pressure? What is its unit of measure?
15. What is relative humidity? What is its range of measurement?
16. Describe the sensors and their features available in the APDS–9960.

References

1. *APDS–9960 Digital Proximity, Ambient Light, RGB and Gesture Sensor Data Sheet,* Avago Technologies, AV02–4191EN, 2015.
2. Arduino homepage, www.arduino.cc
3. *Arduino Nano 33 BLE Sense Product Reference Manual,* ABX00031, 2022.
4. Bluetooth, www.bluetooth.com
5. *Bluetooth Document: 16–bit UUID Numbers Document,* www.bluetooth.com, 2015.
6. Franklin Miller, Jr., *College Physics*, 4th edition, Harcourt, Brace, Jovanovich, 1977.
7. *HTS221–Capacitive digital sensor for relative humidity and temperature,* DocID026333 Rev 4, STMicroelectronics,2016.
8. *LPS22HB– MEMS nano pressure sensor: 260–1260 hPa absolute digital output barometer,* DocID027083 Rev 6 1, STMicroelectronics, 2017.
9. *LSM9DS1– iNEMO inertial module: 3D accelerometer, 3D gyroscope, 3D magnetometer,* DocID025715 Rev 3, STMicroelectronics, 2015.
10. *MP34DT05– MEMS audio sensor omnidirectional digital microphone data sheet,* STMicroelectronics, 2019.
11. *nRF52840 Product Specification,* 4413_417 v1.1, Nordic Semiconductor, 2019.
12. *nRF52840 Advanced multi–protocol System–on–Chip Supporting: Bluetooth low energy (Bluetooth 5), ANT/ANT+, 802.15.4 and 2.4GHz proprietary*, Nordic Semiconductor.

Arduino Nano 33 BLE Sense Power and Interfacing 3

Objectives: After reading this chapter, the reader should be able to:

- Specify a power supply system for an Arduino–based system;
- Describe the voltage and current input/output parameters for the Arduino Nano 33 BLE Sense;
- Apply the voltage and current input/output parameters toward properly interfacing input and output devices to the Nano 33 processing board;
- Interface selected input and output devices to Nano 33 processing board; and
- Describe how to control the speed and direction of a DC motor.

3.1 Overview

We begin this chapter with exploring the power source requirements for the Arduino Nano 33 BLE Sense. We examine how to probably provide power from several sources. The remainder of the chapter provides information on how to interface selected input, output, and high power DC devices to the Nano processor. As mentioned throughout the text, **the Nano 33 the BLE Sense is a 3.3 VDC processor. Inputs to the processor must not exceed this value.**[1]

---

[1] Selected portions of this chapter have been adapted for the Nano 33 BLE Sense 3.3 VDC processor from "Arduino I: Getting Started" for completeness.

© The Author(s), under exclusive license to Springer Nature Switzerland AG 2023
S. F. Barrett, *Arduino V: Machine Learning*, Synthesis Lectures on Digital Circuits
& Systems, https://doi.org/10.1007/978-3-031-21877-4_3

3.2   Arduino Power Requirements

Arduino processing boards may be powered from the USB port during project development. It is highly recommended that an external power supply be employed any time a peripheral component is connected. This will allow developing projects beyond the limited current capability of the USB port.

For the Nano 33 BLE Sense board external regulated VDC voltages from 5 to 18 voltages may be applied to the VIN (pin 15) Power In pin. Ensure the power supply ground is also connected to one the ground pins on the Nano 33 (pins 14 or 19) (www.arduino.cc).

3.3   Voltage Regulators

Voltage regulators provide a stabilized, fixed output voltage as the input voltage varies. A common positive voltage regulator is the 78XX series. The "XX" specifies the regulator output voltage (e.g. 5, 9, 12, etc.). This regulator series has a current rating up to 1.5 Amps. The input voltage to the regulator typically needs to be several volts higher than the desired output voltage (uA7800).

Figure 3.1 provides sample circuits to provide a +5 VDC and a ±5 VDC portable battery source.

3.3.1   Powering the Nano 33 From Batteries

As previously mentioned, for the Nano 33, Arduino recommends a power supply from 5–18 VDC (www.arduino.cc). For low power applications a single 9 VDC battery and clip may be used as shown in Fig. 3.2. For higher power applications, a AA battery pack may be used. The battery supply is provided to a regulator (e.g. 7805) and then to the Nano 33 VIN pin.

3.4   Interfacing Concepts

In the remainder of the chapter we discuss methods to properly interface select peripheral devices to the Nano 33 BLE Sense. The list of interface rules we abide by are short but important. Violating any of the rules may damage the microcontroller, the peripheral, or lead to erratic processor operation. Here are the rules:

- The Nano 33 BLE Sense is a 3.3 VDC processor.
- Inputs (digital and analog) to the processor must not exceed the 3.3 VDC value.
- The maximum current available from an output pin is 10 mA.

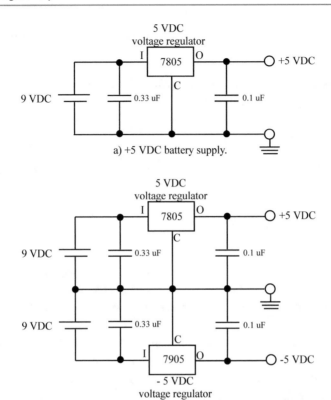

a) +5 VDC battery supply.

b) +/-5 VDC battery supply.

Fig. 3.1 Battery supply circuits employing a 9 VDC battery with a 5 VDC regulators

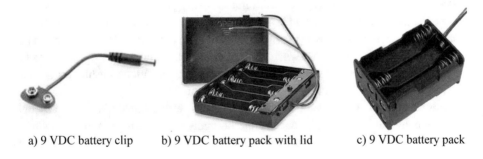

a) 9 VDC battery clip   b) 9 VDC battery pack with lid   c) 9 VDC battery pack

Fig. 3.2 Arduino 9 VDC battery power

- It is highly recommended that an external power supply be employed any time a peripheral component is connected.

With the list of rules in place, we investigate selected input and output devices.

3.5   Input Devices

In this section we describe how to interface several input devices to the Arduino Nano 33 BLE Sense microcontroller.

3.5.1   Switches

Switches come in a variety of types. As a system designer it is up to you to choose the appropriate switch for a specific application. Switch varieties commonly used in microcontroller applications are illustrated in Fig. 3.3a. Here is a brief summary of the different types:

- **Slide switch:** A slide switch has two different positions: on and off. The switch is manually moved to one position or the other. For microcontroller applications, slide switches are available that fit in the profile of a common integrated circuit size dual inline package (DIP). A bank of four or eight DIP switches in a single package is commonly available. Slide switches are used to select specific parameters at system startup.
- **Momentary contact pushbutton switch:** A momentary contact pushbutton switch comes in two varieties: normally closed (NC) and normally open (NO). A normally open switch, as its name implies, does not normally provide an electrical connection between its contacts. When the pushbutton portion of the switch is depressed, the connection between the two switch contacts is made. The connection is held as long as the switch is depressed. When the switch is released the connection is opened. The opposite is true for a normally closed switch. For microcontroller applications, pushbutton switches are available in a small tact type switch configuration.
- **Push on/push off switches:** These switch types are also available in a normally open or normally closed configuration. For the normally open configuration, the switch is depressed to make connection between the two switch contacts. The pushbutton must be depressed again to release the connection.
- **Hexadecimal rotary switches:** Small profile rotary switches are available for microcontroller applications. These switches commonly have sixteen rotary switch positions. As the switch is rotated to each position, a unique four bit binary code is provided at the switch contacts.

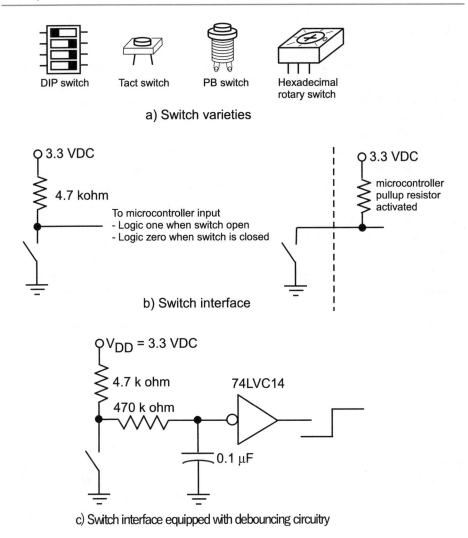

a) Switch varieties

b) Switch interface

c) Switch interface equipped with debouncing circuitry

Fig. 3.3 Switch interface

A common switch interface is shown in Fig. 3.3b. This interface allows a logic one or zero to be properly introduced to a microcontroller input port pin. The basic interface consists of the switch in series with a current limiting resistor. The node between the switch and the resistor is provided to the microcontroller input pin. In the configuration shown, the resistor pulls the microcontroller input up to the supply voltage V_{DD} of 3.3 VDC. When the switch is closed, the node is grounded and a logic zero is provided to the microcontroller input pin. To reverse the logic of the switch configuration, the position of the resistor and the switch is simply reversed.

3.5.1.1 Switch Debouncing

Mechanical switches do not make a clean transition from one position (on) to another (off). When a switch is moved from one position to another, it makes and breaks contact multiple times. This activity may go on for tens of milliseconds. A microcontroller is relatively fast as compared to the action of the switch. Therefore, the microcontroller is able to recognize each switch bounce as a separate and erroneous transition.

To correct the switch bounce phenomena additional external hardware components may be used or software techniques may be employed. A hardware debounce circuit is illustrated in Fig. 3.3c. The node between the switch and the limiting resistor of the basic switch circuit is fed to a low pass filter (LPF) formed by the 470 kΩ resistor and the capacitor. The LPF prevents abrupt changes (bounces) in the input signal from the microcontroller. The LPF is followed by a 74LVC14 Schmitt Trigger, which is simply an inverter equipped with hysteresis. This further limits the switch bouncing.

Switches may also be debounced using software techniques. This is accomplished by inserting a 30 to 50 ms lockout delay in the function responding to port pin changes. The delay prevents the microcontroller from responding to the multiple switch transitions related to bouncing.

You must carefully analyze a given design to determine if hardware or software switch debouncing techniques will be used. It is important to remember that all switches exhibit bounce phenomena and, therefore, must be debounced. An example is provided in the Arduino IDE under Examples –> Digital –> Debounce.

3.6   Output Devices

An external device should not be connected to a microcontroller without first performing careful interface analysis to ensure the voltage, current, and timing requirements of the microcontroller and the external device. In this section, we describe interface considerations for selected external devices. We begin with the interface for a single light emitting diode.

3.6.1   Light Emitting Diodes (LEDs)

An LED is typically used as a logic indicator to inform the presence of a logic one or a logic zero at a specific pin of a microcontroller. An LED has two leads: the anode or positive lead and the cathode or negative lead. To properly bias an LED, the anode lead must be biased at a level approximately 1.7–2.2 volts higher than the cathode lead. This specification is known as the forward voltage (V_f) of the LED. The LED current must also be limited to a safe level known as the forward current (I_f). The diode voltage and current specifications are usually provided by the manufacturer.

Fig. 3.4 LED display device

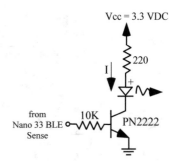

An example of an LED biasing circuit is provided in Fig. 3.4. A logic one is provided by the microcontroller to the base of a low power NPN transistor (e.g. PN2222 or MPQ2222) through a 10 kΩ base resistor. The transistor's emitter lead is grounded. Connected to the transistor's collector lead is a 220 Ω resistor in series with the LED to a 3.3 VDC power supply. The 220 Ω resistor (R) limits the current through the LED. A proper resistor value can be calculated using $R = (V_{cc} - V_{DIODE})/I_{DIODE}$.

3.6.2   Serial Liquid Crystal Display (LCD)

An LCD is an output device to display text information. LCDs come in a wide variety of configurations including multi–character, multi–line format. A 16 x 2 LCD format is common. That is, it has the capability of displaying two lines of 16 characters each. Each display character and line has a specific associated address. The characters are sent to the LCD via American Standard Code for Information Interchange (ASCII) format a single character at a time. The interface between the Nano 33 BLE Sense and a serial LCD was discussed in detail in Chap. 2 for the USART, SPI, and I2C serial communication systems.

3.7   Motor Control Concepts

In this section we investigate different types of DC motors. The motors may be used for locomotion, sensor positioning and scanning, or actuator positioning. For each type of motor we provide a basic theory of operation, microcontroller interface techniques, and example applications. Some of the types of motors available for different applications are shown in Fig. 3.5.

- **DC motor:** A DC motor has a positive and negative terminal. When a DC power supply of suitable voltage and current rating is applied to the motor it will rotate. If the polarity of the supply is switched with reference to the motor terminals, the motor will rotate

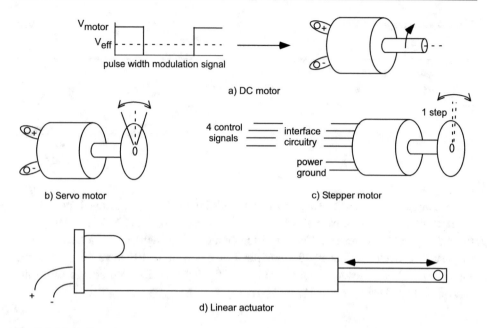

Fig. 3.5 Motor types

in the opposite direction. The speed of the motor is roughly proportional to the applied voltage up to the rated voltage of the motor.

- **Servo motor:** A servo motor provides a precision angular rotation for an applied pulse width modulation (PWM) duty cycle. As the duty cycle of the applied signal is varied, the angular displacement of the motor also varies. This allows the motor to be used in applications to change mechanical positions such as the steering angle of a wheel, position or scan a sensor, or as the main drive motor for a small robot.
- **Stepper motor:** A stepper motor, as its name implies, provides an incremental step change in rotation (typically 2.5 degrees per step) for a step change in control signal sequence. The motor is typically controlled by a two or four wire interface. For the four wire stepper motor, the microcontroller provides a four bit control sequence to rotate the motor clockwise. To turn the motor counterclockwise, the control sequence is reversed. The low power microcontroller control signals are interfaced to the motor via MOSFETs or power transistors to provide for the proper voltage and current requirements of the pulse sequence. The stepper motor may be used to position or scan robot sensors.
- **Linear actuator** The linear actuator is actually a rotating DC motor equipped with gears to translate rotational motion to linear motion. The linear actuator is used when repeatable linear motion, both push and pull, is required. They may be used in robots for steering or for sensor placement.

With this brief overview of motors complete, let's take a closer look at controlling DC motor speed. Methods for controlling other types of motors are provided in "Arduino I: Getting Started."

3.7.1   DC Motor

A direct current or DC motor is typically used in robot applications as the main source of locomotion. The power source for the DC motor is usually an onboard battery supply carried by the robot. We discuss the choice of battery supplies in the next chapter.

A DC motor has a positive and negative terminal. When a DC power supply of suitable voltage and current rating is applied to the motor it will mechanically rotate. If the polarity of the supply is switched with reference to the motor terminals, the motor will rotate in the opposite direction. The speed of the motor is roughly proportional to the applied voltage up to the rated voltage of the motor.

3.7.1.1 DC Motor Ratings
A DC motor is rated using the following common parameters. The requirements of the robot are used to select an appropriate motor [3, 4].

- **voltage:** The maximum operating voltage of a motor is specified in DC volts. At this voltage the motor rotates at its maximum speed specified in revolutions per minute or RPM.
- **current:** When rotating, a motor will draw current from the DC supply. The current, measured in amperes or amps, drawn depends on the load the motor is experiencing. DC motors have the following currents specified: starting current, no load operating current, and stall current. When a motor interface is designed it must withstand these different current values.
  - **start current:** The start current is a surge current that occurs when a motor first starts. The surge is due to the motor overcoming mechanical inertia.
  - **no load current:** The no load operating current is the value of current drawn when the motor is supplied its rated voltage and is not under a mechanical load.
  - **stall current:** The stall current is the current drawn from the supply when the motor is stalled.
- **speed:** The rotational speed of a motor is specified in revolutions per minute or RPM.
- **torque:** Torque is the angular force delivered by the motor at a radius from the motor shaft. It is measured in MKS units of Newton–meters.
- **speed versus torque:** DC motors have several different types of configurations: shunt, compound, and series. Generally speaking, as the motor shaft speed is decreased (e.g. via gears), the motor torque is increased.

- **efficiency:** A DC motor converts DC electrical power into mechanical power. Ideally, we desire all of the electrical power to convert to mechanical power representing 100 percent efficiency. Efficiency is defined as the ratio of mechanical power to electrical power.
- **gears:** Gears are used to reduce the shaft speed of a DC motor while increasing motor torque. Many DC motors are equipped with a gearbox for this purpose.

3.7.1.2 Unidirectional DC Motor Control

In this section we describe the interface between the low power Arduino Nano 33 BLE Sense microcontroller and a DC motor. Recall a digital output pin of the Nano 33 BLE Sense is limited to 3.3 VDC with a current limit of 10 mA.

These voltage and current values are not sufficient to directly drive any type of motor. Therefore, an interface is required to boost the voltage and current to values consistent with a given motor specification. Two common methods of interface include using a Darlington transistor configuration or a power metal–oxide–semiconductor field–effect transistor (MOSFET).

NPN Darlington Transistor

A Darlington configuration consists of two transistors configured as shown in Fig. 3.6a. The emitter of the first transistor is connected to the base of the second transistor. This configuration provides for high current gain. When used as a motor interface, the low output current control signal from the Arduino is boosted to a current value suitable to drive a motor as shown in (b). A series of silicon diodes (1N4001), each with a voltage drop of 0.7 VDC, is used to drop the power supply voltage down to the motor supply voltage rating. Additionally, a reverse biased protection diode is placed across the diode string and motor [2, 5].

Example: The Dagu Magician is a popular, low–cost, two platform robot. The robot is equipped with two drive motors rated at 6 VDC with a maximum current rating of 400 mA and a stall current of 1.5 amp. In this application, the robot motors are powered from an external 9 VDC power supply via an umbilical cable.

We assume a Nano 33 output voltage of 3.3 VDC with a maximum output current of approximately 10 milliamps (we use 5 mA). With this information, the base resistor value for the Darlington transistor is calculated to be approximately 380 Ωs.

$$-V_{OH} + (I_B \times R_B) + (2 \times V_{BE}) = 0$$

$$-3.3 + (0.005 \times R_B) + (2 \times 0.7) = 0$$

$$380 \; Ohms$$

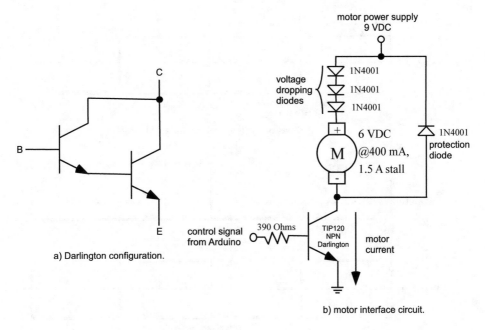

Fig. 3.6 **a** Darlington configuration and **b** motor control circuit

We use a close standard resistor value of 390 Ωs as shown in Fig. 3.6b. The NPN Darlington transistor (TIP 120) boosts the base current to a collector current value suitable to supply the motor. A separate interface circuit is required for each motor.

3.7.1.3 DC Motor Speed Control–Pulse Width Modulation (PWM)

As previously mentioned, DC motor speed may be varied by changing the applied DC voltage. However, PWM control signal techniques may be combined with a motor interface to precisely control the motor speed. With a PWM control signal, a fixed frequency and duty cycle is provided to the motor interface. As shown in Fig. 3.7 the duty cycle of the PWM signal will also be the percentage of the motor supply voltage, or effective DC voltage, applied to the motor and hence the percentage of rated full speed at which the motor will rotate.

The Nano 33 BLE Sense is equipped with five pulse width modulation (PWM) channels. PWM output can be provided on digital pins 1–13 and analog pins A0–A7. We limit use to pins 21, 23, 24, 27, and 28 as shown in the pinout diagram at Fig. 2.2. The Nano 33 BLE sense baseline frequency of the PWM signal is set at 500 Hz.

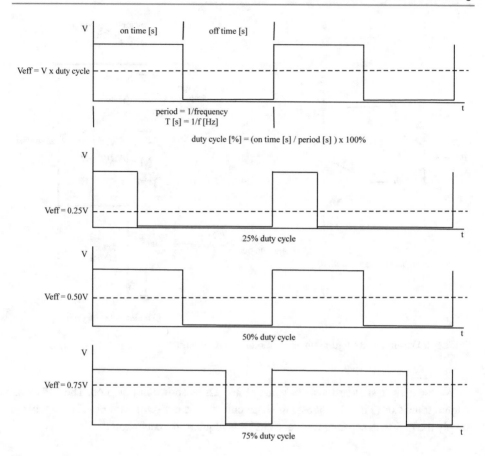

Fig. 3.7 Pulse width modulation

3.8   Application: Dagu Magician Robot

An autonomous, maze navigating robot is equipped with sensors to detect the presence of maze walls and navigate about the maze.[2] The robot has no prior knowledge about the maze configuration. It uses the sensors and an onboard algorithm to determine the robot's next move. The overall goal is to navigate from the starting point of the maze to the end point as quickly as possible without bumping into maze walls as shown in Fig. 3.8. Maze walls are usually painted white to provide a good, light reflective surface; whereas, the maze floor is painted matte black to minimize light reflections.

---

[2] This example appeared in "Arduino I: Getting Started," S. Barrett, Morgan and Claypool Publishers, 2020. It has been adapted with permission for the 3.3 VDC Arduino Nano 33 BLE Sense microcontroller.

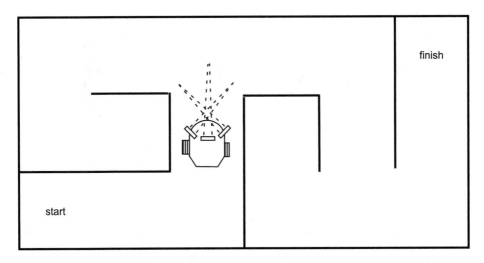

Fig. 3.8 Autonomous robot within maze

It would be helpful to review the fundamentals of robot steering and motor control. Figure 3.9 illustrates the fundamental concepts. Robot steering is dependent upon the number of powered wheels and whether the wheels are equipped with unidirectional or bidirectional control. Additional robot steering configurations are possible. An H–bridge is typically required for bidirectional control of a DC motor. In this example we use unidirectional motor control.

We equip the Dagu Magician Robot (DG007) with an Arduino Nano 33 BLE Sense microcontroller for maze navigation. **Note:** The Nano 33 is a 3.3 VDC microcontroller. Care must be taken to ensure microcontroller inputs do not exceed 3.3 VDC and peripheral output device interfaces accommodate the 3.3 VDC output rated at a maximum of 10 mA DC current per input/output pin. Reference Fig. 3.10. The robot is controlled by two 6 VDC motors which independently drive a left and right wheel. A third non–powered ball provides tripod stability for the robot. We also equip the platform with three Sharp GP2Y0A21YK0F IR sensors as shown in Fig. 3.11. The sensors are available from SparkFun Electronics (www.sparkfun.com). We mount the sensors on a bracket constructed from thin aluminum. Dimensions for the bracket are provided in the figure. Alternatively, the IR sensors may be mounted to the robot platform using "L" brackets available from a local hardware store.

The characteristics of the sensor are provided in Fig. 3.12a. Note that the sensor provides the same output voltage for two different ranges. To remove this ambiguity, we mount the sensor bracket at the front of the robot facing forward such that the left sensor is monitoring maze wall presence on the right of the robot and vice versa. The ambiguity is resolved due to the sensor placement. That is, there is no way for the left or right sensor to be closer than 5 cm right or left maze wall respectively. Hence there is only a single range value for a given sensor output voltage as shown in Fig. 3.12b.

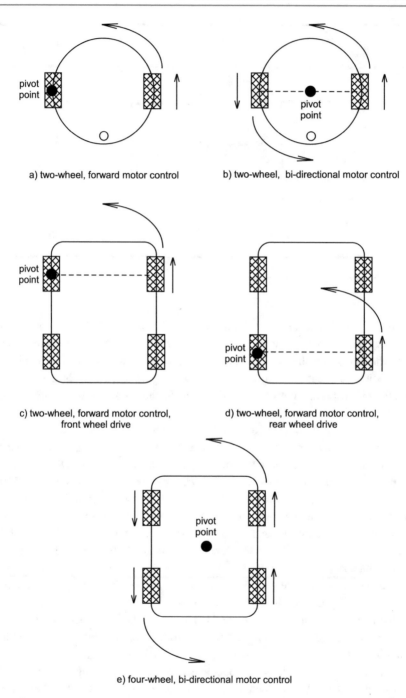

a) two-wheel, forward motor control

b) two-wheel, bi-directional motor control

c) two-wheel, forward motor control,
front wheel drive

d) two-wheel, forward motor control,
rear wheel drive

e) four-wheel, bi-directional motor control

Fig. 3.9 Robot control configurations

IR sensor array

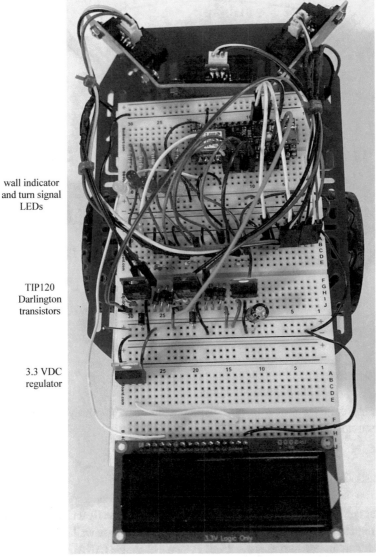

Arduino
Nano 33
BLE Sense

wall indicator
and turn signal
LEDs

TIP120
Darlington
transistors

5 VDC
regulator

3.3 VDC
regulator

Sparkfun LCD-16397

Fig. 3.10 Dagu Magician robot

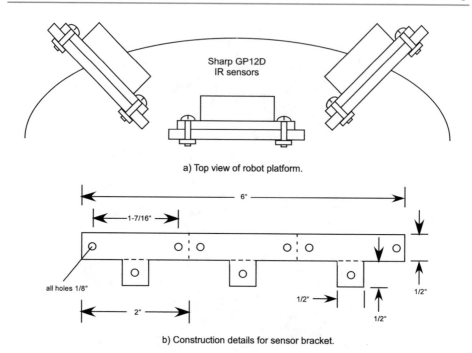

a) Top view of robot platform.

b) Construction details for sensor bracket.

Fig. 3.11 Dagu Magician robot platform modified with three IR sensors

3.8.1   Requirements

The requirements for this project are simple, the robot must autonomously navigate through the maze without touching maze walls.

3.8.2   Circuit Diagram

The circuit diagram for the robot is provided in Fig. 3.13. The three IR sensors (left, middle, and right) are mounted on the leading edge of the robot to detect maze walls. The output from the sensors is fed to three Arduino Nano 33 BLE Sense channels (ANALOG IN 0–2). The robot motors will be driven by pulse width modulated (PWM) channels (PWM: DIGITAL 9 and PWM: DIGITAL 10).

The Arduino Nano 33 BLE Sense is interfaced to the motors via a Darlington NPN transistor (TIP120) with enough drive capability to handle the maximum current requirements of the motor (1.5 A stall current). The robot is powered by a 9 VDC power supply. Three 1N4001 diodes are placed in series with the motor to reduce the motor supply voltage to approximately 6.9 VDC. The 9 VDC supply is also fed to a 5 VDC voltage regulator to power the Arduino Nano 33 via the V_{IN} pin (pin 15). To save on battery expense, it is

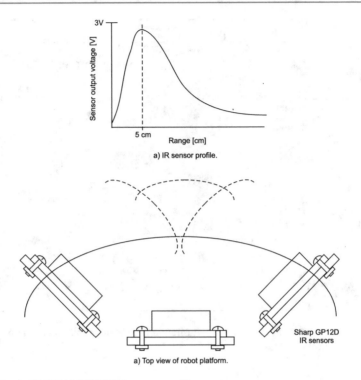

a) IR sensor profile.

a) Top view of robot platform.

Fig. 3.12 Sharp GP2Y0A21YKOF IR sensor profile

recommended to use a 9 VDC, 2A rated inexpensive, wall–mount power supply to provide power to the onboard 5 VDC voltage regulator. A power umbilical of braided wire may be used to provide power to the robot while navigating about the maze.

Structure chart: The structure chart for the robot project is provided in Fig. 3.14.

UML activity diagrams: The UML activity diagram for the robot is provided in Fig. 3.15.

3.8.3 Dagu Magician Robot Control Algorithm

In this section, we provide the basic framework for the robot control algorithm. The control algorithm will read the IR sensors attached to the Arduino Nano 33 ANALOG IN (pins 0–2). In response to the wall placement detected, the algorithm will render signals to turn the robot to avoid the maze walls. Provided in Fig. 3.16 is a truth table that shows all possibilities of maze placement that the robot might encounter. A detected wall is represented with a logic one. An asserted motor action is also represented with a logic one. As previously mentioned, due to the physical placement of the sensor array on the trailing edge of the robot, the sensor detecting maze walls to the right of the robot is physically located on the left side of the robot and vice versa.

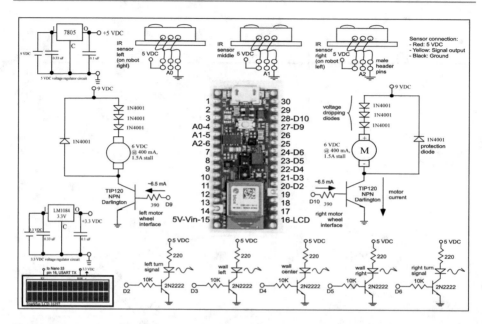

Fig. 3.13 Robot circuit diagram. (Nano 33 illustration used with permission of the Arduino Team (CC BY–NC–SA) www.arduino.cc)

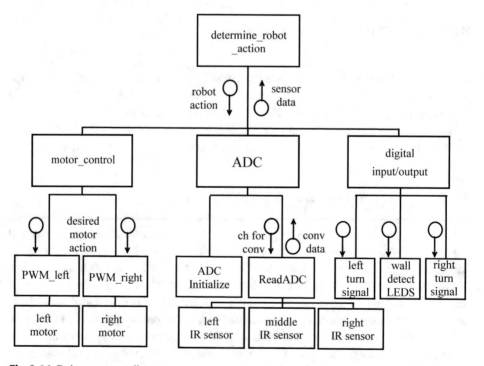

Fig. 3.14 Robot structure diagram

Fig. 3.15 Robot UML activity
diagram

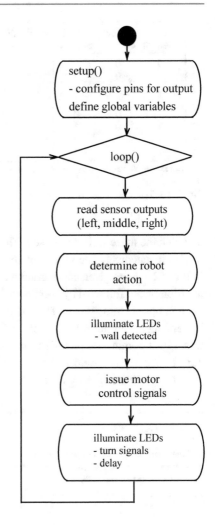

Given the interface circuit used, the robot motors may only be moved in the forward direction. To render a left turn, the left motor is stopped and the right motor is asserted until the robot completes the turn. To render a right turn, the opposite action is required.

The task in writing the control algorithm is to take the UML activity diagram provided in Fig. 3.15 and the actions specified in the robot action truth table Fig. 3.16 and transform both into an Arduino sketch. This may seem formidable but we take it a step at a time.

The control algorithm begins with Arduino Nano 33 pin definitions. Variables are then declared for the readings from the three IR sensors. The two required Arduino functions follow: setup() and loop(). In the setup() function, Arduino Nano 33 pins are declared as output. The loop() begins by reading the current value of the three IR sensors. The 512 value corresponds to a particular IR sensor range. This value may be adjusted to change the range

	Left Sensor	Middle Sensor	Right Sensor	Wall Left	Wall Middle	Wall Right	Left Motor	Right Motor	Left Signal	Right Signal	Comments
0	0	0	0	0	0	0	1	1	0	0	Forward
1	0	0	1	0	0	1	1	1	0	0	Forward
2	0	1	0	0	1	0	1	0	0	1	Right
3	0	1	1	0	1	1	0	1	1	0	Left
4	1	0	0	1	0	0	1	1	0	0	Forward
5	1	0	1	1	0	1	1	1	0	0	Forward
6	1	1	0	1	1	0	1	0	0	1	Right
7	1	1	1	1	1	1	1	0	0	1	Right

Note: The left wall detecting sensor is mounted on the right side of the robot and vice versa.

Fig. 3.16 Truth table for robot action

at which the maze wall is detected. The read of the IR sensors is followed by an eight part if–else if statement. The statement contains a part for each row of the truth table provided in Fig. 3.16. For a given configuration of sensed walls, the appropriate wall detection LEDs are illuminated followed by commands to activate the motors (analogWrite) and illuminate the appropriate turn signals. The analogWrite command issues a signal from 0 to 3.3 VDC by sending a constant from 0 to 255 using pulse width modulation (PWM) techniques. The turn signal commands provide to actions: the appropriate turns signals are flashed and a 1.5 s total delay is provided. This provides the robot 1.5 s to render a turn. This delay may need to be adjusted during the testing phase.

```
//****************************************************************
//Nano_33_robot
//
//The robot is equipped with a Sparkfun LCD-16397
//- Nano 33 BLE Sense, USART TX pin 16 is connected to
//   LCD USART RX pin
//- Provide 3.3 VDC power to the LCD
//****************************************************************

//analog input pins
#define left_IR_sensor    A0      //analog pin A0 (pin 4)-left IR sensor
#define center_IR_sensor A1       //analog pin A1 (pin 5)-ctr IR sensor
#define right_IR_sensor  A2       //analog pin A2 (pin 6)-right IR sensor

                                  //digital output pins
                                  //LED indicators - wall detectors
#define wall_left         3       //digital pin - wall_left
#define wall_center       4       //digital pin - wall_center
#define wall_right        5       //digital pin - wall_right

                                  //LED indicators - turn signals
#define left_turn_signal  2       //digital pin - left_turn_signal
#define right_turn_signal 6       //digital pin - right_turn_signal
```

```
                                   //motor outputs
#define left_motor        9        //digital pin - left_motor
#define right_motor      10        //digital pin - right_motor

int left_IR_sensor_value;          //variable for left IR sensor
int center_IR_sensor_value;        //variable for center IR sensor
int right_IR_sensor_value;         //variable for right IR sensor

void setup()
{
Serial1.begin(9600);               //Baud rate: 9600 Baud
delay(500);                        //Delay for display

                                   //LED indicators - wall detectors
pinMode(wall_left,    OUTPUT);     //configure pin 1 for digital output
pinMode(wall_center,  OUTPUT);     //configure pin 2 for digital output
pinMode(wall_right,   OUTPUT);     //configure pin 3 for digital output

                                    //LED indicators - turn signals
pinMode(left_turn_signal,OUTPUT);  //configure pin 0 for digital output
pinMode(right_turn_signal,OUTPUT); //configure pin 4 for digital output

                                   //motor outputs Â- PWM
pinMode(left_motor,   OUTPUT);     //configure pin 11 for digital output
pinMode(right_motor,  OUTPUT);     //configure pin 10 for digital output
}


void loop()
{
//read analog output from IR sensors
left_IR_sensor_value  = analogRead(left_IR_sensor);   //pin 4 A0
center_IR_sensor_value= analogRead(center_IR_sensor); //pin 5 A1
right_IR_sensor_value = analogRead(right_IR_sensor);  //pin 6 A2

//Clear LCD
//Cursor to line one, character one
Serial1.write(254);                     //Command prefix
Serial1.write(128);                     //Command

//clear display
Serial1.write("                ");
Serial1.write("                ");

//Cursor to line one, character one
Serial1.write(254);                     //Command prefix
Serial1.write(128);                     //Command
Serial1.write("Left  Ctr  Right");
delay(50);

Serial1.write(254);                     //Command to LCD
delay(5);
Serial1.write(192);                     //Cursor line 2, position 1
```

```
delay(5);
Serial1.print(left_IR_sensor_value);
delay(5);
Serial1.write(254);                    //Command to LCD
delay(5);
Serial1.write(198);                    //Cursor line 2, position 8
delay(5);
Serial1.print(center_IR_sensor_value);
delay(5);
Serial1.write(254);                    //Command to LCD
delay(5);
Serial1.write(203);                    //Cursor line 2, position 13
delay(5);
Serial1.print(right_IR_sensor_value);
delay(5);
delay(500);

    //robot action table row 0
    if((left_IR_sensor_value < 512)&&(center_IR_sensor_value < 512)&&
       (right_IR_sensor_value < 512))
      {
                                                //wall detection LEDs
      digitalWrite(wall_left,    LOW);          //turn LED off
      digitalWrite(wall_center, LOW);           //turn LED off
      digitalWrite(wall_right,   LOW);          //turn LED off
                                                //motor control
      analogWrite(left_motor,   128);           //0 (off) to 255 (full speed)
      analogWrite(right_motor, 128);            //0 (off) to 255 (full speed)
                                                //turn signals
      digitalWrite(left_turn_signal,  LOW);     //turn LED off
      digitalWrite(right_turn_signal, LOW);     //turn LED off
      delay(500);                               //delay 500 ms
      digitalWrite(left_turn_signal,  LOW);     //turn LED off
      digitalWrite(right_turn_signal, LOW);     //turn LED off
      delay(500);                               //delay 500 ms
      digitalWrite(left_turn_signal,  LOW);     //turn LED off
      digitalWrite(right_turn_signal, LOW);     //turn LED off
      delay(500);                               //delay 500 ms
      digitalWrite(left_turn_signal,  LOW);     //turn LED off
      digitalWrite(right_turn_signal, LOW);     //turn LED off
      analogWrite(left_motor, 0);               //turn motor off
      analogWrite(right_motor,0);               //turn motor off
      }

    //robot action table row 1
    else if((left_IR_sensor_value < 512)&&(center_IR_sensor_value < 512)&&
            (right_IR_sensor_value > 512))
      {
                                                //wall detection LEDs
      digitalWrite(wall_left,    LOW);          //turn LED off
      digitalWrite(wall_center, LOW);           //turn LED off
      digitalWrite(wall_right,   HIGH);         //turn LED on
                                                //motor control
```

```
   analogWrite(left_motor,  128);          //0(off) to 255(full)
   analogWrite(right_motor, 128);          //0(off) to 255(full)
                                           //turn signals
   digitalWrite(left_turn_signal,  LOW);   //turn LED off
   digitalWrite(right_turn_signal, LOW);   //turn LED off
   delay(500);                             //delay 500 ms
   digitalWrite(left_turn_signal,  LOW);   //turn LED off
   digitalWrite(right_turn_signal, LOW);   //turn LED off
   delay(500);                             //delay 500 ms
   digitalWrite(left_turn_signal,  LOW);   //turn LED off
   digitalWrite(right_turn_signal, LOW);   //turn LED off
   delay(500);                             //delay 500 ms
   digitalWrite(left_turn_signal,  LOW);   //turn LED off
   digitalWrite(right_turn_signal, LOW);   //turn LED off
   analogWrite(left_motor, 0);             //turn motor off
   analogWrite(right_motor,0);             //turn motor off
   }

//robot action table row 2
else if((left_IR_sensor_value < 512)&&(center_IR_sensor_value > 512)&&
        (right_IR_sensor_value < 512))
   {
                                           //wall detection LEDs
   digitalWrite(wall_left,   LOW);         //turn LED off
   digitalWrite(wall_center, HIGH);        //turn LED on
   digitalWrite(wall_right,  LOW);         //turn LED off
                                           //motor control
   analogWrite(left_motor,  128);          //0(off) to 255 (full)
   analogWrite(right_motor, 0);            //0(off) to 255 (full)
                                           //turn signals
   digitalWrite(left_turn_signal,  LOW);   //turn LED off
   digitalWrite(right_turn_signal, HIGH);  //turn LED on
   delay(500);                             //delay 500 ms
   digitalWrite(left_turn_signal,  LOW);   //turn LED off
   digitalWrite(right_turn_signal, LOW);   //turn LED off
   delay(500);                             //delay 500 ms
   digitalWrite(left_turn_signal,  LOW);   //turn LED off
   digitalWrite(right_turn_signal, HIGH);  //turn LED on
   delay(500);                             //delay 500 ms
   digitalWrite(left_turn_signal,  LOW);   //turn LED off
   digitalWrite(right_turn_signal, LOW);   //turn LED off
   analogWrite(left_motor, 0);             //turn motor off
   analogWrite(right_motor,0);             //turn motor off
   }

//robot action table row 3
else if((left_IR_sensor_value < 512)&&(center_IR_sensor_value > 512)&&
        (right_IR_sensor_value > 512))
   {
                                           //wall detection LEDs
   digitalWrite(wall_left,   LOW);         //turn LED off
   digitalWrite(wall_center, HIGH);        //turn LED on
   digitalWrite(wall_right,  HIGH);        //turn LED on
```

```
                                          //motor control
    analogWrite(left_motor,  0);          //0(off) to 255 (full)
    analogWrite(right_motor, 128);        //0(off) to 255 (full)
                                          //turn signals
    digitalWrite(left_turn_signal,  HIGH); //turn LED on
    digitalWrite(right_turn_signal, LOW);  //turn LED off
    delay(500);                            //delay 500 ms
    digitalWrite(left_turn_signal,  LOW);  //turn LED off
    digitalWrite(right_turn_signal, LOW);  //turn LED off
    delay(500);                            //delay 500 ms
    digitalWrite(left_turn_signal,  HIGH); //turn LED on
    digitalWrite(right_turn_signal, LOW);  //turn LED off
    delay(500);                            //delay 500 ms
    digitalWrite(left_turn_signal,  LOW);  //turn LED off
    digitalWrite(right_turn_signal, LOW);  //turn LED off
    analogWrite(left_motor, 0);            //turn motor off
    analogWrite(right_motor,0);            //turn motor off
    }

//robot action table row 4
else if((left_IR_sensor_value > 512)&&(center_IR_sensor_value < 512)&&
        (right_IR_sensor_value < 512))
    {
                                          //wall detection LEDs
    digitalWrite(wall_left,   HIGH);      //turn LED on
    digitalWrite(wall_center, LOW);       //turn LED off
    digitalWrite(wall_right,  LOW);       //turn LED off
                                          //motor control
    analogWrite(left_motor,  128);        //0(off) to 255 (full)
    analogWrite(right_motor, 128);        //0(off) to 255 (full)
                                          //turn signals
    digitalWrite(left_turn_signal,  LOW);  //turn LED off
    digitalWrite(right_turn_signal, LOW);  //turn LED off
    delay(500);                            //delay 500 ms
    digitalWrite(left_turn_signal,  LOW);  //turn LED off
    digitalWrite(right_turn_signal, LOW);  //turn LED off
    delay(500);                            //delay 500 ms
    digitalWrite(left_turn_signal,  LOW);  //turn LED off
    digitalWrite(right_turn_signal, LOW);  //turn LED off
    delay(500);                            //delay 500 ms
    digitalWrite(left_turn_signal,  LOW);  //turn LED off
    digitalWrite(right_turn_signal, LOW);  //turn LED off
    analogWrite(left_motor, 0);            //turn motor off
    analogWrite(right_motor,0);            //turn motor off
    }

//robot action table row 5
else if((left_IR_sensor_value > 512)&&(center_IR_sensor_value < 512)&&
        (right_IR_sensor_value > 512))
    {
                                          //wall detection LEDs
    digitalWrite(wall_left,   HIGH);      //turn LED on
    digitalWrite(wall_center, LOW);       //turn LED off
```

```
    digitalWrite(wall_right,   HIGH);          //turn LED on
                                               //motor control
    analogWrite(left_motor,   128);            //0(off) to 255 (full)
    analogWrite(right_motor, 128);             //0(off) to 255 (full)
                                               //turn signals
    digitalWrite(left_turn_signal,  LOW);      //turn LED off
    digitalWrite(right_turn_signal, LOW);      //turn LED off
    delay(500);                                //delay 500 ms
    digitalWrite(left_turn_signal,  LOW);      //turn LED off
    digitalWrite(right_turn_signal, LOW);      //turn LED off
    delay(500);                                //delay 500 ms
    digitalWrite(left_turn_signal,  LOW);      //turn LED off
    digitalWrite(right_turn_signal, LOW);      //turn LED off
    delay(500);                                //delay 500 ms
    digitalWrite(left_turn_signal,  LOW);      //turn LED off
    digitalWrite(right_turn_signal, LOW);      //turn LED off
    analogWrite(left_motor, 0);                //turn motor off
    analogWrite(right_motor,0);                //turn motor off
    }

//robot action table row 6
else if((left_IR_sensor_value > 512)&&(center_IR_sensor_value > 512)&&
        (right_IR_sensor_value < 512))
    {
                                               //wall detection LEDs
    digitalWrite(wall_left,    HIGH);          //turn LED on
    digitalWrite(wall_center, HIGH);           //turn LED on
    digitalWrite(wall_right,   LOW);           //turn LED off
                                               //motor control
    analogWrite(left_motor,   128);            //0(off) to 255 (full)
    analogWrite(right_motor,  0);              //0(off) to 255 (full)
                                               //turn signals
    digitalWrite(left_turn_signal,  LOW);      //turn LED off
    digitalWrite(right_turn_signal, HIGH);     //turn LED on
    delay(500);                                //delay 500 ms
    digitalWrite(left_turn_signal,  LOW);      //turn LED off
    digitalWrite(right_turn_signal, LOW);      //turn LED off
    delay(500);                                //delay 500 ms
    digitalWrite(left_turn_signal,  LOW);      //turn LED off
    digitalWrite(right_turn_signal, HIGH);     //turn LED off
    delay(500);                                //delay 500 ms
    digitalWrite(left_turn_signal,  LOW);      //turn LED OFF
    digitalWrite(right_turn_signal, LOW);      //turn LED OFF
    analogWrite(left_motor, 0);                //turn motor off
    analogWrite(right_motor,0);                //turn motor off
    }

//robot action table row 7
else if((left_IR_sensor_value > 512)&&(center_IR_sensor_value > 512)&&
        (right_IR_sensor_value > 512))
    {
                                               //wall detection LEDs
    digitalWrite(wall_left,    HIGH);          //turn LED on
```

```
digitalWrite(wall_center, HIGH);        //turn LED on
digitalWrite(wall_right,  HIGH);        //turn LED on
                                        //motor control
analogWrite(left_motor,  128);          //0(off) to 255 (full)
analogWrite(right_motor, 0);            //0(off) to 255 (full)
                                        //turn signals
digitalWrite(left_turn_signal,  LOW);   //turn LED off
digitalWrite(right_turn_signal, HIGH);  //turn LED on
delay(500);                             //delay 500 ms
digitalWrite(left_turn_signal,  LOW);   //turn LED off
digitalWrite(right_turn_signal, LOW);   //turn LED off
delay(500);                             //delay 500 ms
digitalWrite(left_turn_signal,  LOW);   //turn LED off
digitalWrite(right_turn_signal, HIGH);  //turn LED on
delay(500);                             //delay 500 ms
digitalWrite(left_turn_signal,  LOW);   //turn LED off
digitalWrite(right_turn_signal, LOW);   //turn LED off
analogWrite(left_motor, 0);             //turn motor off
analogWrite(right_motor,0);             //turn motor off
}
}


//*********************************************************************
```

3.8.4   Testing the Control Algorithm

It is recommended that the algorithm be first tested without the entire robot platform. This may be accomplished by connecting the three IR sensors and LEDS to the appropriate pins on the Arduino Nano 33 as specified in Fig. 3.13. In place of the two motors and their interface circuits, two LEDs with the required interface circuitry may be used. The LEDs will illuminate to indicate the motors would be on during different test scenarios. Once this algorithm is fully tested in this fashion, the Arduino Nano 33 may be mounted to the robot platform and connected to the motors. Full up testing in the maze may commence. Enjoy!

3.9   Summary

We began this chapter with exploring the power source requirements for the Arduino Nano 33 BLE Sense. We then examined how to probably provide power from several sources. The remainder of the chapter provided information on how to interface selected input, output, and high power DC devices to the Nano processor.

3.10 Problems

1. List the interface guidelines for the Arduino Nano 33 BLE Sense.
2. What will happen if a microcontroller is used outside of its prescribed operating envelope?
3. What are the power requirements for the Arduino Nano 33 BLE Sense? Describe options for providing power to the microcontroller.
4. What is the role and function of a voltage regulator?
5. Design a regulated battery pack for the Arduino Nano 33 BLE Sense. Specify all components. How long will the battery pack you designed power the Arduino Nano 33 BLE Sense.
6. Can an LED with a series limiting resistor be directly driven by the Nano 33 BLE Sense microcontroller? Explain.
7. What is switch bounce? Describe two techniques to minimize switch bounce. What happens if a switch is not debounced.
8. What function within the Arduino IDE provides for PWM? Explain all required variables for the function.

References

1. Arduino homepage, www.arduino.cc
2. Boylestad, R. and L. Nashelsky, (1982). *Electronic Devices and Circuit Theory*, third edition, Prentice–Hall.
3. Clark D. and M. Owings, (2003). *Building Robot Drive Trains*, McGraw Hill.
4. Fitzgerald, A., C. Kingsley, and S. Umans, (2003). *Electric Machinery*, sixth edition, McGraw Hill.
5. Sedra, A and K. Smith, (2004). *Microelectronic Circuits*, fifth edition, Oxford University Press.
6. *TIP31C Power Transistors (NPN)*, ST Microelectronics, st.com, 2006.
7. *TIP32C Power Transistors (PNP)*, ST Microelectronics, st.com, 2006.
8. *TIP120, TIP121, TIP122 (NPN); TIP125, TIP126, TIP127 (PNP) plastic medium–power complementary silicon transistors (TIP120D)*, ON Semiconductor, onsemi.com, 2007.

Artificial Intelligence and Machine Learning

4

Objectives: After reading this chapter, the reader should be able to:

- Describe the relationship between the concepts of artificial intelligence, machine learning, and deep learning;
- Provide a list of applications within the categories of artificial intelligence, machine learning, and deep learning;
- Sketch a timeline of key developments within the world of artificial intelligence;
- Provide a summary of the theory supporting a K nearest neighbor (KNN) application;
- Design and implement a KNN application on an Arduino processor.
- Provide a summary of the theory supporting decision tree design and analysis; and
- Design and implement a decision tree on an Arduino processor.

4.1 Overview

With the introduction to the Arduino IDE (Chap. 1), the Nano 33 BLE Sense (Chap. 2), and interface techniques (Chap. 3) complete; we concentrate on Artificial Intelligence (AI) and Machine Learning (ML) concepts and applications for microcontroller–based systems for the remainder of the book.

In a recent release, the Arduino Team stated "Arduino is on a mission to make machine learning simple enough for everyone to use [https://www.blog.arduino.cc]." Those acquainted with AI and ML concepts might counter these concepts are most appropriate for more powerful computing platforms. However, recent developments have allowed certain AI and ML applications to be executed on microcontrollers once they have been trained. We will see the training task, in certain applications, may also be accomplished on a micro-controller. Furthermore, we explore applications that lend themselves to remote, battery operated microcontroller–based AI and ML applications [4]. In the remainder of the book

S. F. Barrett, *Arduino V: Machine Learning*, Synthesis Lectures on Digital Circuits & Systems, https://doi.org/10.1007/978-3-031-21877-4_4

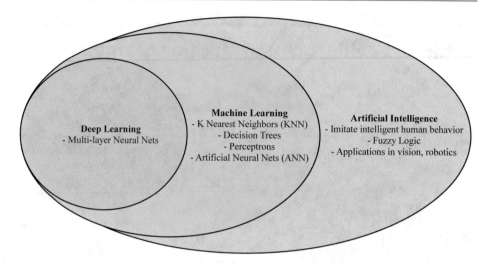

Fig. 4.1 Artificial intelligence and machine learning (Mueller and Massaron)

we limit our discussions to AI and ML techniques specifically for microcontrollers. The intent is to introduce the concepts and allow you to practice on low cost, accessible Arduino hardware and software.

Figure 4.1 illustrates the relationship between Artificial Intelligence, Machine Learning, and Deep Learning. The goal of Artificial Intelligence is for computing machinery to imitate and mimic intelligent human behavior. Some trace the origins of AI back to 1300 BCE [4]. We limit our historical review to AI developments within the 20th century and forward. Following a brief historical review, this chapter explores the concept of K Nearest Neighbors (KNN) and Decision Tree classification techniques.

Within the realm of AI, we explore Fuzzy Logic in Chap. 5. The overall goal of fuzzy logic is to control a system using a series of rule statements of the form "IF–THEN." The input conditions of the "IF" statement, the antecedents, are obtained by taking precise, crisp input information from input sensors and transducers and mapping them to fuzzy input linguistic (word) variables. This is called the fuzzification step. The "THEN" portion of the rule, the consequences, are the control commands back to the system. Again, linguistic variables are used to describe the control output. The control output is defuzzified to obtain precise, crisp output control signals. The mapping of inputs to outputs via the "IF–THEN" statements are provided by the system designer.

Machine Learning is a category under the broad umbrella of Artificial Intelligence. Its goal is to develop algorithms to control a process. The developed algorithm undergoes a learning step where input data is used to confirm or develop desired controller outputs. During the learning process the algorithm adjusts certain weights and biases to improve the performance of the algorithm. Within the realm of ML we explore decision trees, K nearest neighbor (KNN) algorithms, perceptrons, artificial neurons, and artificial neural networks

(ANNs). We explore decision trees and KNN applications in this chapter and the remaining topics in Chap. 6.

Deep Learning involves the development of algorithms using multi–layer artificial neural networks (ANN). The concepts provided in Chap. 6 will allow the reader to develop deep learning networks. We conclude Chap. 6 with a brief introduction to advanced AI and ML tools and applications.

4.2 A Brief History of AI and ML Developments

Provided in Fig. 4.2 is a summary of key developments in artificial intelligence and machine learning. Admittedly, many contributions are not on the timeline.[1]

We begin with the work of George Boole in the development of Boolean Algebra to describe the "fundamental laws of those operations of the mind." In the late 1930s Claude Shannons thesis described how to use Boolean Algebra to optimize telephone communications routing. This led to the common use of Boolean Algebra's common use in the design of early computers.

Several years later in the midst of World War II McCulloch and Pitts developed the first mathematical model of the neuron. In 1949 Donald Hebb described how synapses between neurons are strengthened when used repeatedly. In 1956 Dartmouth College hosted the first AI Conference. The conference gathered leading AI scientists for discussions lasting two months. A year later Frank Rosenblatt developed the model and implementation of the single perceptron. We explore this model in some detail later in the book.

In 1959 Arthur Samuel developed the basic concepts of machine learning including supervised and unsupervised learning techniques. In the mid–1960s Alexey Ivakhnenko conducted early work in multi–layered neural networks which was termed "Deep Learning" approximately two decades later. Around the same time, Lofti Zadeh developed the fundamental and mathematical concepts of Fuzzy Logic. We explore Fuzzy Logic in some detail later in the book.

In 1969 Minsky and Papert identified and demonstrated the limitations of the perceptron model in solving nonlinear classification models. This spurred additional work to develop neural networks capable of solving nonlinear classification models. Five years later Paul Werbos provided foundational, theoretical work for backpropagation in his Ph.D. dissertation. In 1986 Rumelhart, McClelland, and other associated researchers developed and applied concepts of machine learning with backpropagation.

We now fast forward to 2005. At this time the Arduino Team released the first Arduino processor based on concepts of open source hardware and software which launched a global movement for accessible computing. Following this same philosophy, in a recent release, the Arduino Team stated "Arduino is on a mission to make machine learning simple enough

---

[1] Please see "Artificial Intelligence An Illustrated History" by Clifford Pickover for a thorough and fascinating treatment of this topic back to 1300 BCE.

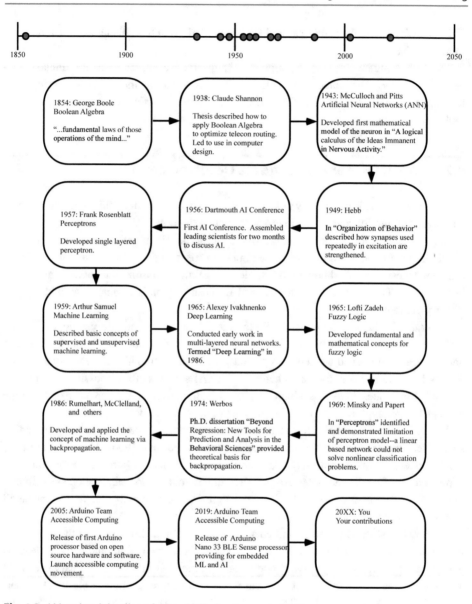

Fig. 4.2 Abbreviated timeline of AI and ML developments [4, 9]

for everyone to use [https://www.blog.arduino.cc]." In 2019 the Arduino Team releases the Nano 33 BLE Sense processor providing for accessible ML and AI.

In the remainder of this book we explore the use of Arduino processors for AI and ML applications. In this chapter we explore the concepts of K Nearest Neighbors (KNN) and Decision Trees. We explore Fuzzy Logic in Chap. 5, Perceptrons and Neural Networks in Chap. 6. We conclude Chap. 6 with a brief introduction to advanced AI and ML deep learning tools and applications.

4.3   K Nearest Neighbors

The overall goal of the K Nearest Neighbors or KNN technique is to classify a new object by comparing its features to an established dataset. It is based on the premise that objects with similar features and a shared classification will cluster together when mapped into a feature space. This technique of classifying objects can be traced back to the work Fix and Hodges in 1951 [6]. The KNN technique is considered a supervised classification technique. That is a KNN algorithm is provided a training set of objects with a known classification before it can be used to classify objects outside of the training set.

For example, in Fig. 4.3 a color classifier is illustrated. The algorithm was trained with a set of ten data points for each desired, known classification (color). In this example, the Red, Green, and Blue (RGB) color component of each sample was stored along with its corresponding classification or tag.

Once trained, a new non–classified object's RGB components are submitted to the algorithm to determine its classification. The RGB components are compared to the nearest "K" neighboring RGB components already in the tagged data set. In the figure the algorithm is using a K value of three to determine the distance to the three closest neighbors. Three common distance measures are shown including the Euclidean distance, the Manhattan distance, and the Minkowski distance measures (IBM).

An Arduino sketch for a KNN based color classifier is provided below. It is adapted from the "ColorClassifer" within the Arduino KNN Library. The UML activity diagram for the sketch is provided in Fig. 4.4. The color classifier uses the ADPS[2] system onboard the Arduino Nano 33 BLE Sense.

---

[2] Digital proximity, ambient light, RGB, and gesture sensor (ADPS–9960 onboard the Arduino Nano 33 BLE Sense.

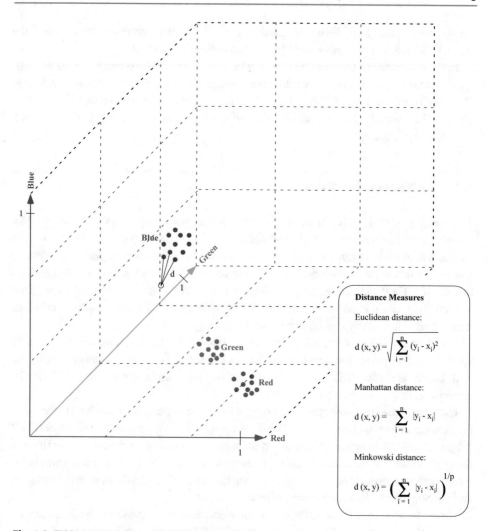

Fig. 4.3 K Nearest Neighbor (KNN) classifier (IBM)

To test the sketch, download the Arduino KNN Library from the Tools − > Manage Library tab within the Arduino IDE. Color samples for red, green, and blue may be constructed from color construction paper. Compile and upload the sketch to the Arduino Nano 33 BLE Sense. Follow the user prompts provided in the Serial Monitor to train and then use the color classifier.

Fig. 4.4 KNN color classifier

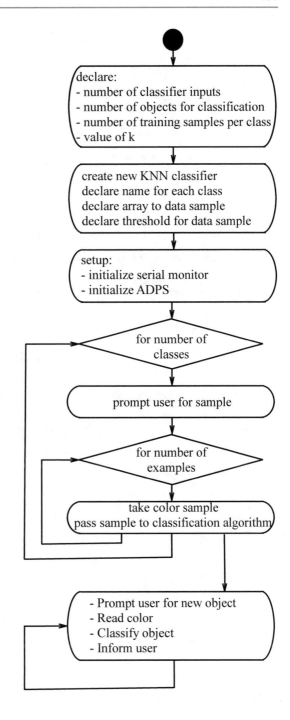

```
//**********************************************************************
//KNN color classification
//
//This sketch classifies objects using a color sensor.
//
//First you 'teach' the Arduino by putting an example of each object
//close to the color sensor.
//
//After the training step, the Arduino will guess the name of objects
//it is shown based on how similar the color is to the examples.
//
//The "k" is adjustable and set to 5.
//
//Processor: Arduino Nano 33 BLE Sense
//
//When compiled and uploaded, open the serial console.
//- Follow prompts to provide sample objects.
//- Once trained, user is prompted for objects.
//- Algorithm will guess the color of the object provided.
//
//Sketch adapted from ColorClassifer from Arduino KNN Library
//
//This example code is in the public domain.
//**********************************************************************

#include <Arduino_KNN.h>
#include <Arduino_APDS9960.h>

const int INPUTS = 3;              //Number of input values passed to classifier.
                                   //Input is color sensor R, G, B level data
const int CLASSES = 3;             //Number of objects for classification
                                   //(e.g. Red, Blue, Green)
const int EXAMPLES_PER_CLASS = 10; //Number of data samples per object
                                   //for training
const int K = 5;                   //number of neighbors considered
float color[INPUTS];               //Data storage array
                                   //Used to pass data to KNN alogorithm
const int THRESHOLD = 5;           //Threshold for color brightness

KNNClassifier myKNN(INPUTS);       //Create  a new KNNClassifier
                                   //Names for each class (obj ect type)
String label[CLASSES] = {"Red", "Blue", "Green"};

void setup()
{
Serial.begin(9600);
while (!Serial);

if(!APDS.begin())                 //Ensure ADPS is present
  {
  Serial.println("Failed to initialize APDS!");
  while (1);
  }

Serial.println("Arduino k-NN color classifier");

//Prompt user for the name of each object
for(int currentClass = 0; currentClass < CLASSES; currentClass++)
  {
  //Prompt user for samples of object
  for(int currentExample = 0; currentExample < EXAMPLES_PER_CLASS; currentExample++)
    {
    Serial.print("Show me an example ");
```

```
    Serial.println(label[currentClass]);
    readColor(color);                       //Wait for an object then read its color
    myKNN.addExample(color, currentClass); //Add example color to the k-NN model
    }
  //Wait for user to move object away
  while(!APDS.proximityAvailable()||APDS.readProximity()==0){}
  }
}

void loop()
{
int classification;

//Wait for the object to move away again
while (!APDS.proximityAvailable() || APDS.readProximity()==0){}

Serial.println("Let me guess your object");
readColor(color);                       //Wait for an object then read its color

classification = myKNN.classify(color, K);//Classify the object

Serial.print("You showed me: ");  //Print the classification
Serial.println(label[classification]);
}

//*******************************************************************

void readColor(float color[])
{
int red, green, blue, proximity, colorTotal = 0;

//Wait for the object to move close
while (!APDS.proximityAvailable() || APDS.readProximity() > 0){}

//Wait until color is bright enough
while (colorTotal < THRESHOLD)
  {
  //Sample if color is available and object is close
  if(APDS.colorAvailable())
    {
    APDS.readColor(red, green, blue);  // Read color and proximity
    colorTotal = (red + green + blue);
    }
  }

//Normalise the color sample data and put it in the classifier input array
color[0] = (float)red   / colorTotal;
color[1] = (float)green/ colorTotal;
color[2] = (float)blue / colorTotal;

//Print the red, green and blue percentage values
Serial.print(color[0]);  Serial.print(",");
Serial.print(color[1]);  Serial.print(",");
Serial.println(color[2]);
}

//*******************************************************************
```

4.4   Decision Trees

Decision Trees are a powerful, graphical method to illustrate a complex decision making process. The decision tree is constructed using supervised learning. That is, a set of observations or objects with accompanying outcomes or classifications are used to systematically construct the decision tree.[3]

The data set consists of a number of attributes or features to describe an object for classification. Each attribute has a series of values to describe the feature. Accompanying the attributes and values is an observed outcome or classification.

The data set to construct the decision tree may be extracted from an existing database or it may be prepared by an expert in the field. Once the decision tree has been constructed it may be used to render decisions or classifications with new observations outside the original data set.

J.R. Quinlan in the classic work "Induction of Decision Trees" provides an example to determine whether or not to play tennis on a given day based on a series of weather attributes and prior decisions made as shown in Fig. 4.5 [5].

Observations	Attributes				Outcome/ Decision
Observation#	Outlook	Temperature	Humidity	Windy?	Play Tennis?
1	sunny	hot	high	F	N
2	sunny	hot	high	T	N
3	overcast	hot	high	F	Y
4	rain	mild	high	F	Y
5	rain	cool	normal	F	Y
6	rain	cool	normal	T	N
7	overcast	cool	normal	T	Y
8	sunny	mild	high	F	N
9	sunny	cool	normal	F	Y
10	rain	mild	normal	F	Y
11	sunny	mild	normal	T	Y
12	overcast	mild	high	T	Y
13	overcast	hot	normal	F	Y
14	rain	mild	high	T	N

Attribute: value 1, value 2, value 3
Outlook: sunny, overcast, rain
Temperature: hot, mild, cool,
Humidity: high, normal
Windy: true (T), false (F)

Fig. 4.5 Data set for tennis play [5]

---

[3] Background information for this section is from [5, 8].

The attributes to describe the day include: outlook, temperature, humidity, and windy. The values for each attribute are provided in the figure. The data set consists of series of 14 observations taken on different Saturdays. For each observation, the observed value of each attribute was recorded and whether or not tennis was played on that day. The goal is to develop a decision tree to process the attribute values on a new day, not in the data set, to determine if tennis will be played or not.

This example may seem trivial. You might counter: "I do not need a decision tree to determine whether to play tennis or not." However, the example serves as an approachable template to learn and implement decision tree fundamentals. Decision trees may be used for a wide variety of applications including weather prediction, medical diagnosis, and robot control.

To begin decision tree construction, statistics of the data set are gathered. These statistics may be manually gathered or determined using the decision tree algorithm. In this example, we use a mixture of both approaches and also a complement of arrays to store the collected information. As shown in Fig. 4.6, the following arrays are used to implement the decision tree:

- dt_array: contains the observation data set,
- attrib_array: contains attributes and the associated values,
- attr_cnt_array: the count for each attribute resulting in play tennis ("Y") and no play ("N"),
- entropy_array: contains the entropy calculation for each attribute, and
- tree_result_array: contains the resulting decision tree.

It is important to note that many different data structures may be used to construct a decision tree. Examples include structures, binary trees, and recursive algorithms. We use a one–dimensional array (tree_result_array) to store the decision tree. The array, although one–dimensional, may be used to store and construct a two–dimensional tree as shown in Fig. 4.7.

The base of the tree is called the root node. Each node contains the name of the attribute and three accompanying value fields (left, center, and right). Each of the fields has a fixed relationship to its children nodes. The fields within the children nodes also have a fixed relationship to its grandchildren nodes. As shown in the figure, the structure of the tree is fixed and the relationship between the root nodes, children nodes, and grandchildren nodes are known and fixed.

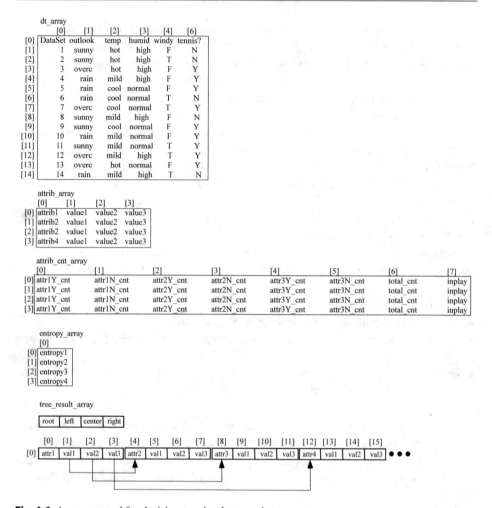

Fig. 4.6 Array set used for decision tree implementation

The Unified Modeling language (UML) activity diagram or flowchart for developing the decision tree algorithm is provided in Fig. 4.8.

The decision tree is assembled step–by–step starting at the root node. The root node is the attribute resulting in the lowest value of entropy (disorder). The entropy calculation for a given attribute is shown in Fig. 4.9. The fields of the root node (left, center, and right) contain the names of the values associated with the root attribute. The count for each attribute value resulting in the play/no play determine is provided in the attribute_count_array.

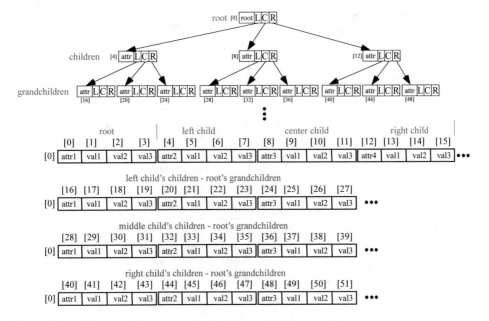

Fig. 4.7 Two–dimensional array used for decision tree implementation

Once the root node is loaded into the result array (tree_result_array); the left, center, and right fields are examined to establish the contents of the children nodes. The children nodes may contain the next available attribute with the lowest entropy or leaf values. The children node fields are then examined to establish the contents of the grandchildren nodes. This process continues until no attributes are left to form the tree. As the decision tree is constructed, each attribute performs a decision step to split data into smaller sets and reduce the disorder of the data set.

The resulting decision tree for playing tennis is provided in Fig. 4.10. Note the attribute for temperature does not appear in the final decision tree. The algorithm developed excluded attributes with a entropy greater than 0.90. Examination of the temperature attribute in the attribute_count_array shows this feature does not impact the decision to play tennis.

The Arduino sketch to implement the decision tree for tennis play follows. It was implemented using the Arduino Nano 33 BLE Sense.

Fig. 4.8 Decision tree UML
activity diagram

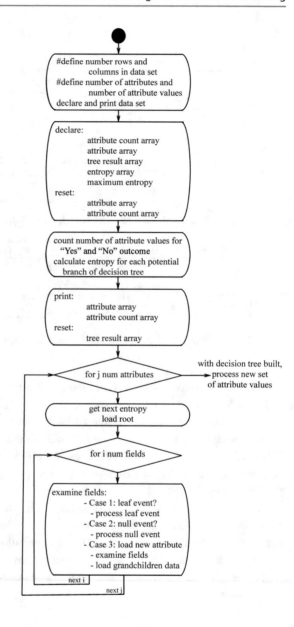

Value 1	Value 2	1st term	2nd term	entropy	entropy total
2	3	0.4	0.6	0.97	
4	0	1	0	0.00	
3	2	0.6	0.4	0.97	0.694
2	2	0.50	0.50	1.00	
4	2	0.67	0.33	0.92	
3	1	0.75	0.25	0.81	0.911
3	4	0.43	0.57	0.99	
6	1	0.86	0.14	0.59	0.788
6	2	0.75	0.25	0.81	
3	3	0.50	0.50	1.00	0.892

$\text{Entropy(value1, value2)} = - \text{value1}/(\text{value1}+\text{value2})*\log_2*\text{value1}/(\text{value1}+\text{value2})$
$\qquad\qquad\qquad\qquad\quad - \text{value2}/(\text{value1}+\text{value2})*\log_2*\text{value2}/(\text{value1}+\text{value2})$

$\text{1st\_term} = \text{value } 1/(\text{value } 1 + \text{value } 2)$

$\text{2nd\_term} = \text{value } 2/(\text{value } 1 + \text{value } 2)$

$\text{Entropy (value 1, value 2)} = - (\text{1st\_term} * \log10(\text{1st\_term})/\log10(2)) - (\text{2nd\_term} * \log10(\text{2nd\_term})/\log10(2))$

Fig. 4.9 Entropy calculations

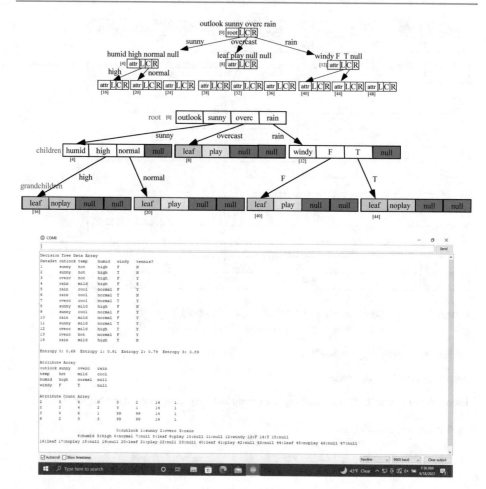

Fig. 4.10 Decision tree for tennis play

```
//****************************************************************
//decision_tree
//****************************************************************

#define num_rows 15
#define num_cols  6
#define max_num_attrib_vals 4
#define num_attrib 4

unsigned int first_attrib = 1, last_attrib = (num_cols - 1);
unsigned int child_incr, gr_child_incr;

const char* dt_array[num_rows][num_cols] =
{{"DataSet", "outlook", "temp",    "humid", "windy", "tennis?"},
 {     "1",   "sunny",  "hot",     "high",    "F",      "N"},
 {     "2",   "sunny",  "hot",     "high",    "T",      "N"},
 {     "3",   "overc",  "hot",     "high",    "F",      "Y"},
```

```
{      "4",    "rain", "mild",    "high",    "F",    "Y"},
{      "5",    "rain", "cool",  "normal",    "F",    "Y"},
{      "6",    "rain", "cool",  "normal",    "T",    "N"},
{      "7",   "overc", "cool",  "normal",    "T",    "Y"},
{      "8",  "sunny", "mild",    "high",    "F",    "N"},
{      "9",  "sunny", "cool",  "normal",    "F",    "Y"},
{     "10",    "rain", "mild",  "normal",    "F",    "Y"},
{     "11",  "sunny", "mild",  "normal",    "T",    "Y"},
{     "12",   "overc", "mild",    "high",    "T",    "Y"},
{     "13",   "overc",  "hot",  "normal",    "F",    "Y"},
{     "14",    "rain", "mild",    "high",    "T",    "N"}};

unsigned int attr_cnt_array[num_attrib][(max_num_attrib_vals * 2)];
const char* attrib_array[num_attrib][max_num_attrib_vals];
const char* tree_result_array[num_attrib * 4 * 4];
double entropy_array[num_attrib][1];
double entropy_total, entropy_val;
double entropy_max = 0.90;


void setup()
{
Serial.begin(9600);
while(!Serial);
print_DT_array();
reset_attrib_array();
reset_attr_cnt_array();
num_values_per_attrib();
print_attrib_array();
print_attr_cnt_array();
reset_tree_result_array();
build_decision_tree();
print_tree_result_array();
}


void loop()
{
//navigate decision tree with new values of attribute values


}

//********************************************************************

void print_DT_array(void)
{
unsigned int i,j;

while(!Serial);
Serial.println("Decision Tree Data Array");
for(i=0; i< num_rows; i++)                    //row selector
   {
   for(j=0; j< num_cols; j++)                 //column selector
      {
      Serial.print(dt_array[i][j]);
      Serial.print("\t");
      }
   Serial.println("");
   }
Serial.println("");
}


//********************************************************************

void num_values_per_attrib(void)
{
const char* value1;
const char* value2;
```

```
const char* value3;
unsigned int val2_set, val3_set;
unsigned int attr1Y_cnt, attr2Y_cnt, attr3Y_cnt;
unsigned int attr1N_cnt, attr2N_cnt, attr3N_cnt;
unsigned int total_attr_cnt;
unsigned int i, j, p;

total_attr_cnt = 0;
for(j=first_attrib; j< last_attrib; j++)            //col selector
  {                                                 //set attr
  value1 = "null"; value2 = "null"; value3 = "null";
  value1 = dt_array[1][j];                          //set 1st attr
  val2_set = 0; val3_set = 0;                       //reset flags
  total_attr_cnt = 0;
  for(i=2; i< num_rows; i++)
    {                                               //set 2nd attr
    if((value1 != dt_array [i][j])&&(val2_set == 0))
      {
      value2 = dt_array [i][j];
      val2_set = 1;
      }                                             //set 3rd attr
    else if ((value1!=dt_array [i][j])&&(value2!=dt_array [i][j])&&(val3_set == 0))
      {
      value3 = dt_array [i][j];
      val3_set = 1;
      }
    else
      {
      ;
      }
    }

  //store attribute values to attrib_array
  attrib_array[j-1][0] = dt_array[0][j];
  attrib_array[j-1][1] = value1;
  attrib_array[j-1][2] = value2;
  attrib_array[j-1][3] = value3;

  attr1Y_cnt = 0; attr2Y_cnt = 0; attr3Y_cnt = 0;  //reset attr cnt
  attr1N_cnt = 0; attr2N_cnt = 0; attr3N_cnt = 0;  //reset attr cnt
  for(i=1; i< num_rows; i++)
    {
    //count number of attrb in each category
    if((value1 == dt_array[i][j])&&(dt_array[i][num_cols - 1] == "Y"))
      {attr1Y_cnt++;}
    else if ((value1 == dt_array[i][j])&&(dt_array[i][num_cols - 1] == "N"))
      {attr1N_cnt++;}
    else if((value2 == dt_array[i][j])&&(dt_array[i][num_cols - 1] == "Y"))
      {attr2Y_cnt++;}
    else if ((value2 == dt_array[i][j])&&(dt_array[i][num_cols - 1] == "N"))
      {attr2N_cnt++;}
    else if((value3 == dt_array[i][j])&&(dt_array[i][num_cols - 1] == "Y"))
      {attr3Y_cnt++;}
    else if ((value3 == dt_array[i][j])&&(dt_array[i][num_cols - 1] == "N"))
      {attr3N_cnt++;}
    else
      {
      ;
      }
    }
    //end for i

  //store results to attr_cnt_array
  if((attr1Y_cnt + attr1N_cnt) != 0)
    {
    attr_cnt_array[j-1][0] = attr1Y_cnt;
    attr_cnt_array[j-1][1] = attr1N_cnt;
```

```
      total_attr_cnt = total_attr_cnt + attr1Y_cnt + attr1N_cnt;
      }
    else
      {
      attr_cnt_array[j-1][0] = 99;
      attr_cnt_array[j-1][1] = 99;
      }
    if((attr2Y_cnt + attr2N_cnt) != 0)
      {
      attr_cnt_array[j-1][2] = attr2Y_cnt;
      attr_cnt_array[j-1][3] = attr2N_cnt;
      total_attr_cnt = total_attr_cnt + attr2Y_cnt + attr2N_cnt;
      }
    else
      {
      attr_cnt_array[j-1][2] = 99;
      attr_cnt_array[j-1][3] = 99;
      }
    if((attr3Y_cnt + attr3N_cnt) != 0)
      {
      attr_cnt_array[j-1][4] = attr3Y_cnt;
      attr_cnt_array[j-1][5] = attr3N_cnt;
      total_attr_cnt = total_attr_cnt + attr3Y_cnt + attr3N_cnt;
      }
    else
      {
      attr_cnt_array[j-1][4] = 99;
      attr_cnt_array[j-1][5] = 99;
      }
    attr_cnt_array[j-1][6] =  attr_cnt_array[j-1][6] + total_attr_cnt;
    attr_cnt_array[j-1][7] = 1;   //indicates attribute still in play

    //calculate entropy of attribute and attribute values
    entropy_total = 0;
    for(p = 0; p < ((max_num_attrib_vals-1) * 2); p=p+2)
      {
      if(attr_cnt_array[j-1][p] != 99)
        {
        entropy_val = calculate_entropy(attr_cnt_array[j-1][p],
                                attr_cnt_array[j-1][p+1]);
        entropy_val = (((double)((attr_cnt_array[j-1][p])) + (double)(attr_cnt_array[j-1][p+1]))/
                        ((double)((attr_cnt_array[j-1][6])))) * entropy_val;
        entropy_total = entropy_total + entropy_val;
        }
      }
      entropy_array[j-1][0] = entropy_total;
    }//end for j
  print_entropy_array();
delay(100);
}

//************************************************************************

void print_attrib_array(void)
{
unsigned int k, m;

while(!Serial);
Serial.println("Atrribute Array");
for(k=0; k < num_attrib; k++)                    //row selector
  {
  for(m=0; m < max_num_attrib_vals; m++)         //column selector
    {
    Serial.print(attrib_array[k][m]);
    delay(10);
    Serial.print("\t");
    delay(10);
```

```
     }
   Serial.println("");
   }
Serial.println("");
}

//************************************************************************

void reset_attrib_array(void)
{
unsigned int k, m;

for(k=0;  k < num_attrib; k++)                        //row selector
   {
   for(m=0; m < max_num_attrib_vals; m++)             //column selector
      {
      attrib_array[k][m] = "null";
      }
   }
}

//************************************************************************

void print_attr_cnt_array(void)
{
unsigned int k,m;

Serial.println("Atrribute Count Array");
for(k=0; k < (num_attrib); k++)                       //row selector
   {
   for(m=0; m < (max_num_attrib_vals * 2); m++) //col selector
      {
      Serial.print(attr_cnt_array[k][m]);
      Serial.print("\t");
      }
      Serial.println("");
   }
Serial.println("");
}

//************************************************************************

void reset_attr_cnt_array(void)
{
unsigned int k,m;

for(k=0; k < (num_attrib); k++)                       //row selector
   {
   for(m=0; m < (max_num_attrib_vals * 2); m++) //col selector
      {
      attr_cnt_array[k][m] = 0;
      }
   }
}

//************************************************************************

double calculate_entropy(unsigned int first_val, unsigned int second_val)
{
double entropy_val;
double first_val_dbl, second_val_dbl;
double first_val_dbl_sub, second_val_dbl_sub;

first_val_dbl = (double)(first_val);
second_val_dbl = (double)(second_val);

//prevents NaN in log calculation
```

```
if(first_val_dbl < 1.0)
  first_val_dbl_sub = 0.005;
else
  first_val_dbl_sub = first_val_dbl;

if(second_val_dbl < 1.0)
  second_val_dbl_sub = 0.005;
else
  second_val_dbl_sub = second_val_dbl;

entropy_val = - ((first_val_dbl/(first_val_dbl + second_val_dbl))  *
  (log10((first_val_dbl_sub/(first_val_dbl_sub+second_val_dbl_sub))))/log10(2))-
  ((second_val_dbl/(first_val_dbl+second_val_dbl))*
  (log10((second_val_dbl_sub/(first_val_dbl_sub+second_val_dbl_sub))))/log10(2));

return entropy_val;
}

//**********************************************************************

void print_entropy_array(void)
{
unsigned int k;

for(k=0; k < (num_attrib); k++)                    //row selector
  {
  Serial.print("Entropy ");
  Serial.print(k);
  Serial.print(": ");
  Serial.print(entropy_array[k][0]);
  Serial.print("   ");
  }
Serial.println(" ");
Serial.println(" ");
}


//**********************************************************************
//void build_decision_tree(void)
//**********************************************************************

void build_decision_tree(void)
{
unsigned int i,j,k;
unsigned int attr_found;              //provides row# for attr array
unsigned int new_attr_found;          //provides row# for attr array
unsigned int array_counter = 0;       //position in tree_result_array

for(j=0; j < num_attrib; j++)             //find entropy values in order
  {                                       //lowest to highest
  attr_found = get_next_entropy();        //provides row# attrib_array
  if((attr_found != 99)&&(entropy_array[attr_found][0]<=entropy_max))
                                          //reports 99 if no new value
    {                                     //load attrib to tree_result_array
    tree_result_array[array_counter]=attrib_array[attr_found][0];//attr name
    tree_result_array[array_counter+1]=attrib_array[attr_found][1];//attr val1
    tree_result_array[array_counter+2]=attrib_array[attr_found][2];//attr val2
    tree_result_array[array_counter+3] = attrib_array[attr_found][3]; //attr val3

    for(i=0; i <= 2; i++)
      {
      //Case 1: examine for leaf event
      child_incr =((i+1)*4);
      if(((attr_cnt_array[attr_found][(2*i)])==0)||((attr_cnt_array[attr_found][(2*i)+1])==0))
        {
        //Set leaf values of child nodes
        tree_result_array[array_counter+child_incr+0] = "leaf";
```

```
    if((attr_cnt_array[attr_found][2*i]) > (attr_cnt_array[attr_found][(2*i)+1]))
       {
       tree_result_array[array_counter+child_incr+1]  = "play";
       tree_result_array[array_counter+child_incr+2]  = "null";
       tree_result_array[array_counter+child_incr+3]  = "null";
       }
    else
       {
       tree_result_array[array_counter+child_incr+1] = "no play";
       tree_result_array[array_counter+child_incr+2] = "null";
       tree_result_array[array_counter+child_incr+3] = "null";
       }
    }
//Case 2: null encountered - load null
else if((tree_result_array[array_counter+1]) == "null")
    {
    tree_result_array[array_counter+child_incr+0] = "null";   //attr name
    tree_result_array[array_counter+child_incr+1] = "null";   //attr val1
    tree_result_array[array_counter+child_incr+2] = "null";   //attr val2
    tree_result_array[array_counter+child_incr+3] = "null";   //attr val3
    }
else
  {
  //Case 3: load new attribute
  new_attr_found = get_next_entropy();      //provides row# attrib_array
  if((new_attr_found != 99)&&(entropy_array[new_attr_found][0]<=entropy_max))
     {
     //load attrib to tree_result_array
     tree_result_array[array_counter+child_incr+0] = attrib_array[new_attr_found][0]; //attr name
     tree_result_array[array_counter+child_incr+1] = attrib_array[new_attr_found][1]; //attr val1
     tree_result_array[array_counter+child_incr+2] = attrib_array[new_attr_found][2]; //attr val2
     tree_result_array[array_counter+child_incr+3] = attrib_array[new_attr_found][3]; //attr val3

     //determine grandchild increment
     if(child_incr == 4)                       //left child
       gr_child_incr = child_incr + 12;
     else if (child_incr == 8)                 //center child
       gr_child_incr = child_incr + 20;
     else if (child_incr == 12)                //right child
       gr_child_incr = child_incr + 28;

     for(k=0; k<=2; k++)
       {
       //load grand child leaves
       if(((attr_cnt_array[new_attr_found][2*k])!=99)&&
          ((attr_cnt_array[new_attr_found][(2*k)+1])!=99)&&
          (tree_result_array[array_counter+child_incr+0] != "leaf")&&
          (tree_result_array[array_counter+child_incr+k] != "null"))
          {
          tree_result_array[array_counter+gr_child_incr+0+(k*4)] = "leaf";
          if((attr_cnt_array[new_attr_found][2*k]) > (attr_cnt_array[new_attr_found][(2*k)+1]))
             {
             tree_result_array[array_counter+gr_child_incr+1+(k*4)] = "play";
             tree_result_array[array_counter+gr_child_incr+2+(k*4)] = "null";
             tree_result_array[array_counter+gr_child_incr+3+(k*4)] = "null";
             }
          else
             {
             tree_result_array[array_counter+gr_child_incr+1+(k*4)] = "noplay";
             tree_result_array[array_counter+gr_child_incr+2+(k*4)] = "null";
             tree_result_array[array_counter+gr_child_incr+3+(k*4)] = "null";
             }
          }
       }
     }
  else
     {
```

```
            //load 'null"
            tree_result_array[array_counter+child_incr+0] = "null";  //attr name
            tree_result_array[array_counter+child_incr+1] = "null";  //attr val1
            tree_result_array[array_counter+child_incr+2] = "null";  //attr val2
            tree_result_array[array_counter+child_incr+3] = "null";  //attr val3
            }
        }
     }//end for(i...
    }//end if(attrb_found != 99)

  else //if(attrb_found != 99)
    {
    ;
    }
  array_counter = array_counter + 16;
  }//end for (j..
}

//*********************************************************************

unsigned int get_next_entropy(void)
{
unsigned int n;

double low_entropy = 1.0;
unsigned int row_found;
unsigned int found_one = 0;
unsigned int lowest_row_found = 10;

for(n=0; n <= num_attrib; n++)
  {
  if((entropy_array[n][0] < low_entropy)&&(attr_cnt_array[n][7] == 1))
    {
    low_entropy = entropy_array[n][0];
    lowest_row_found = n;
    found_one = 1;
    }
  }//end for
  attr_cnt_array[lowest_row_found][7] = 0;
  row_found = lowest_row_found;

if(found_one == 0) row_found = 99;

return row_found;
}

//*********************************************************************

void print_tree_result_array(void)
{
unsigned int i;

Serial.print("\t\t\t");
for(i=0; i< (num_attrib * 4 * 4); i++)
  {
  if(tree_result_array[i] != "x")
    {
    Serial.print(i);
    Serial.print(":");
    Serial.print(tree_result_array[i]);
    Serial.print(" ");
    }
  if((i==3)||(i==15)||(i==47))
    {
    Serial.println(" ");
    if(i==3)
      Serial.print("\t\t");
```

```
      }
    }
Serial.println(" ");
}

//*********************************************************************

void reset_tree_result_array(void)
{
unsigned int i;

for(i=0; i< (num_attrib * 4 * 4); i++)
    {
    tree_result_array[i] = "x" ;
    }
}

//*********************************************************************
```

Once the tree is assembled, it may be used to process new attribute values outside of the original data set to make a decision using tree traversal techniques. We explore this next in the chapter Applications section.

4.5   Application: KNN Classifier

Earlier in the chapter we explored a KNN color classifier for three colors (Red, Green, and Blue). Expand the example to eight different colors. Explore the tradeoffs of varying the value of "k" and the number samples taken of each color sample.

4.6   Application: Decision Trees

In the previous section we explored the design and implementation of a decision tree. In this section we extend our results.

Activities:

1. The function "print_tree_result_array" provided a basic depiction of the resulting decision tree. Modify this function to depict the interconnections between different tree nodes.
2. Provided below is the "loop" portion of the decision tree code provided in the previous section. The "loop" portion provides for tree traversal and decision making. Combine the two portions of the code and test the tree. Add the "loop" portion of the code to the UML activity diagram provided in Fig. 4.8.
3. Develop a new decision tree using a data set of your choosing.

```
//*********************************************************************

void loop()
{
unsigned int i = 0;
unsigned int i_stop = 0;
int j;

i_stop = 0;
//navigate decision tree with new values of attribute values
while((tree_result_array[i] != "x")&&(i_stop!=99))
   {
   //root node processing
   Serial.println("  ");
   Serial.println("Decision Tree Processing");
   Serial.println(tree_result_array[i]);

   if(tree_result_array[i] != "leaf")
      {
      Serial.println("Select the attribute value:");
      }

   if(tree_result_array[i] != "leaf")
    . {
      Serial.print("1: ");
      Serial.println(tree_result_array[i+1]);
      }
   else
      {
      Serial.println(tree_result_array[i+1]);
      i_stop = 99;
      }

   if(tree_result_array[i+2] != "null")
   //&&(tree_result_array[i] != "leaf"))
      {
      Serial.print("2: ");
      Serial.println(tree_result_array[i+2]);
      }

   if(tree_result_array[i+3] != "null")
   //&&(tree_result_array[i] != "leaf"))
      {
      Serial.print("3: ");
      Serial.println(tree_result_array[i+3]);
      }

   if(i_stop != 99)
      {
      //flush input buffer
      while(Serial.available() >0)
         {
         Serial.read();
```

```
      }
   //request input value from user via serial monitor
   Serial.println("Enter a value, press [Send]");
   while(Serial.available()==0){}          //wait for user input data
   j = Serial.parseInt();

   //Serial.print("i: ");  Serial.print(i);
   //Serial.print("   j: ");  Serial.println(j);
   //children level
   if((j==1)&&(i==0))
      i = 4;                    //left
   else if((j==2)&&(i==0))
      i = 8;                    //center
   else if ((j==3)&&(i==0))
      i = 12;                   //right

   //grandchildren level
   else if((j==1)&&(i==4))
      i = 16;                   //left
   else if((j==2)&&(i==4))
      i = 20;                   //center
   else if((j==3)&&(i==4))
      i = 24;                   //right

   //grandchildren level
   else if((j==1)&&(i==8))
      i_stop = 99;  //28        //left
   else if((j==2)&&(i==8))
      i_stop = 99;  //32        //center
   else if((j==3)&&(i==8))
      i_stop = 99;  //36        //right

   //grandchildren level
   else if((j==1)&&(i==12))
      i = 40;                   //left
   else if((j==2)&&(i==12))
      i = 44;                   //center
   else if((j==3)&&(i==12))
      i_stop = 99;              //right
   else
      i = 0;
   } //i!=99

 }//while
 i = 0;
}//loop

//******************************************************************
```

4.7 Summary

We began this chapter with an overview of Artificial Intelligence and Machine Learning applications and historical developments. We then explored to classification techniques: K Nearest Neighbors and Decision Trees. We concentrate on additional Artificial Intelligence and Machine Learning concepts and applications for microcontroller–based systems for the remainder of the book.

4.8 Problems

1. Choose one innovation in recent AI and ML history. Explore the innovation and write a one–page paper. Provide a brief summary of the innovation's background theory and its main contribution to AI and ML science.
2. In the KNN algorithm what is the significance in the choice of "K"? What is the significance of choosing a smaller or larger value of "K"?
3. In the KNN algorithm how will algorithm respond to a sample it has not been trained for?
4. In the KNN algorithm, what are the tradeoffs of using a smaller or larger training set?
5. Earlier in the chapter we explored a KNN color classifier for three colors (Red, Green, and Blue). Expand the example to eight different colors.
6. What is a decision tree? How is it constructed?
7. What is entropy? What does it represent? How is it calculated?
8. The Decision Tree function "print_tree_result_array" provided a basic depiction of the resulting decision tree. Modify this function to depict the interconnections between different tree nodes.
9. Develop a new decision tree using a data set of your choosing.

References

1. Arduino Team, *Get started with machine learning on Arduino*, https://www.blog.arduino.cc, October 15, 2019.
2. G. Lawton, *Machine Learning on Microcontrollers Enables AI,* https://www.targettech.com, November 17, 2021.
3. J.P. Mueller and L. Massaron, *Artificial Intelligence for Dummies,* John Wiley and Sons, Inc, 2018.
4. C.A. Pickover, *Artificial Intelligence an Illustrated History,* Sterling Publishing Co., Inc., 2019.
5. J.R. Quinlan, *Induction of Decision Trees,* Machine Learning 1: pages 81–106, 1986.
6. B. W. Silverman and M. C. Jones, textitE. Fix and J.L. Hodges (1951): An Important Contribution to Nonparametric Discriminant Analysis and Density Estimation: Commentary on Fix and Hodges (1951), International Statistical Review, Dec. 1989, Vol. 57, No. 3 (Dec. 1989), pp 233–238.
7. AI Computing Comes to Memory Chips, IEEE Spectrum, Jan 2022

8. O. Theobald, *Machine Learning for Absolute Beginners: A Plain English Introduction,* second edition, 2017.

9. B.J. Wythoff, *Backpropagation Neural Networks–A Tutorial,* Chemometrics and Intelligent Laboratory Systems, 18 (1993), 115–155.

Fuzzy Logic

5

Objectives: After reading this chapter, the reader should be able to

- Provide a real world example of a control system;
- Describe the differences between a traditional control system design and one implemented with fuzzy logic techniques;
- Describe the similarities and differences between Boolean two value logic and multi–value fuzzy logic;
- Briefly describe the origins of the fuzzy logic approach to control system design;
- Sketch a diagram and describe the steps of designing a fuzzy logic controller;
- Implement a fuzzy logic controller using the Arduino Embedded Fuzzy Logic Library (eFLL); and
- Design a fuzzy logic control system to control an autonomous maze navigating robot.

5.1   Overview Concepts

This chapter describes how to control a process using fuzzy logic techniques. Up to this point in the Arduino book series we have used Boolean Logic techniques. That is, the digital logic inputs and outputs to the microcontroller have either been at logic one (true) or logic zero (false). In fact, we have gone to great lengths to describe proper peripheral interface techniques to preserve these logic values.

In this chapter we explore fuzzy logic that allows multiple levels of truth between logic one and zero. We find that many real world control problems lend themselves to fuzzy logic implementation. As an example, as we travel, our goal is always to arrive safely at our desired destination. Along the way we must constantly stay alert to ever changing road conditions, other vehicles, and even wildlife and domestic animals. Many vehicles are now equipped

© The Author(s), under exclusive license to Springer Nature Switzerland AG 2023
S. F. Barrett, *Arduino V: Machine Learning*, Synthesis Lectures on Digital Circuits
& Systems, https://doi.org/10.1007/978-3-031-21877-4_5

with control systems to automatically adjust their speed in response to detected obstacles. Obstacles may range from slower moving vehicles or even a large animal.[1]

Intuitively, the system will respond differently for an obstruction that is very close rather than much further away. For example, if a large animal (e.g. an elk) steps out right in front of my vehicle, I would like the control system to rapidly apply strong brake pressure to bring the vehicle to a controlled but abrupt stop. On the other hand, if the control system detects an obstacle much farther away from the car, the brake system may be gently applied to slow the car a little bit. Fuzzy logic allows the design a system of this type. Later in the chapter, we investigate a vehicle control system of this type in some detail.

A traditional control system design is based on developing a system mathematical model. Traditional control systems have been used for decades to control all types of complex systems. In 1965, L. A. Zadeh in the seminal work "Fuzzy Sets," introduced the concept of a "continuum of grades of membership" with membership values ranging from zero to one. He also provided functions typically found in Boolean logic (e.g. AND, OR, etc.). He noted that many human decision thought processes do not use precise mathematical models [11].

In this chapter we investigate the design of fuzzy logic controllers in some detail. We begin with a brief review of the key concepts and design of a fuzzy logic controller. We next explore the Arduino Embedded Fuzzy Logic Library (eFLL). The library was developed by a team (Alves, Lira, Lemos, Kridi, and Leal) of the Robotic Research Group at the State University of Piaui in Tersini, Piaui, Brazil. The library contains features to design complex, fuzzy logic control systems, and also provides several excellent examples that may be used as templates to design other systems (Alves). We explore the examples in some detail. In an earlier writing project, Daniel Pack and I explored an autonomous,[2] maze navigating robot equipped with two IR sensors to detect maze walls. In the Application section, we equip the Dagu Magician Robot from Chap. 3 with an eFLL based fuzzy logic control system.

5.2   Theory

In this section we provide an overview of designing a fuzzy logic based control system. Our goal is to provide a systematic, step–by–step design process. Along the way we introduce related terminology and concepts. We purposely use many figures to illustrate key concepts. References for this section include: [1, 4, 5, 8].

Figure 5.1 provides an overview of the fuzzy control system design process. The overall goal is to control a system by taking precise, crisp input information from input sensors and transducers; map the input information, to precise, crisp output control signals using a series of rule statements of the form "IF–THEN."

---

[1] Several years back on a fishing trip my car collided with a wayward deer on an interstate in Eastern Montana. I am thankful that my passengers and I survived the encounter.

[2] "68HC12 Microcontroller Theory and Application, D. J. Pack and S. F. Barrett, Prentice Hall," 2002.

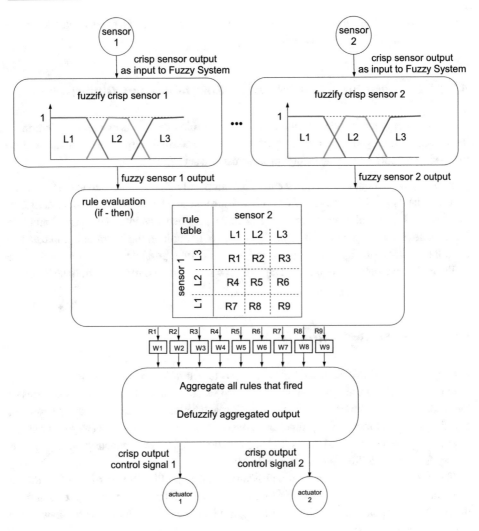

Fig. 5.1 Fuzzy control system design

Along the way, the crisp input signals are fuzzified and converted to linguistic (word) variables. The inputs are then combined using "AND" and "OR" style logic to develop a series of "IF–THEN" style rules. The rules link the input variables to desired output control signals. The rules are developed by an "expert," you, the system designer. There may be multiple rules linking input conditions to desired output control signals. Each rule is evaluated and assigned a firing weight if activated. For a given set of input conditions, multiple rules may fire.

The rules that have fired are then aggregated (combined) to determine an overall fuzzy response. The fuzzy response is then defuzzified to provide crisp output control signals.

Let's take a closer look at each step. We use an extended example of a fuzzy controlled maze following robot to illustrate each step.

5.2.1   Establish Fuzzy Control System Goal, Inputs, and Outputs

Before beginning the detailed design of a fuzzy control system. It is essential to determine the overall system goal, the available inputs, and the desired outputs. The inputs will be mapped to outputs during the design process using a set of rules.

Example: Throughout this section we use an example of a fuzzy controlled maze following robot. In the Application section of Chap. 3 we introduced a Dagu Magician robot platform. We designed a control system consisting of three IR sensors as input peripherals and two DC motors as output peripherals to navigate the robot autonomously through a maze. The overall goal was to navigate through the maze as quickly as possible without touching maze walls. In this example, we equip the Dagu robot with a fuzzy control system to accomplish this same goal.

5.2.2   Fuzzify Crisp Sensor Values

Once system goals have been set, the next step is to determine the fuzzy input membership functions for each of the sensor inputs. The crisp sensor signals are provided by a series of input transducers. A input membership function is developed for each sensor as shown in Fig. 5.1. The input membership functions consist of a series of linguistic (word) variables. The span of the linguistic variables is defined by a trapezoid (or trap) function. Various forms of trap functions are illustrated in Fig. 5.2 (Alves).

The specific trap functions are defined using the sensor profile. The crisp numerical output from the sensor is mapped to a specific linguistic variable. If the crisp numerical output from the sensor corresponds to two different linguistic variables, the linguistic variable with the smaller value is chosen.

Example: In the robot example, we use only two IR sensors to navigate through the maze. To allow the robot to detect obstacles directly in front, a front facing IR sensor is used. Also, a right facing IR sensor is used.

To design the input membership functions for the front and right IR sensor, the IR sensor profile is divided into three different zones: obstacle close, obstacle near, and obstacle far as shown in Fig. 5.3a. The IR sensor profile is used with the output sensor value to construct the input membership functions as shown in Fig. 5.3b. An input membership function is provided for both the front and right facing sensors.

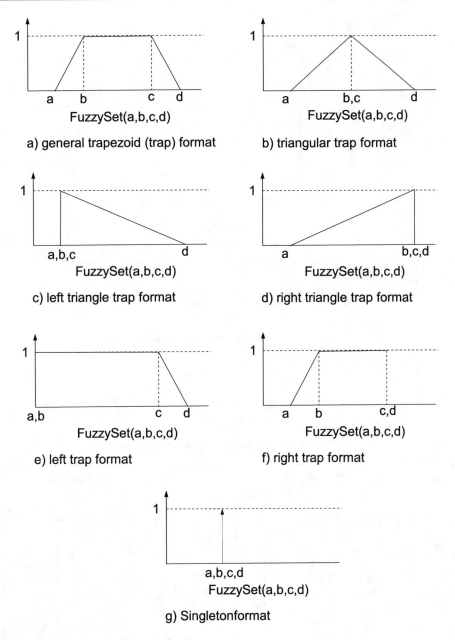

Fig. 5.2 Fuzzy trapezoids (traps) (Alves)

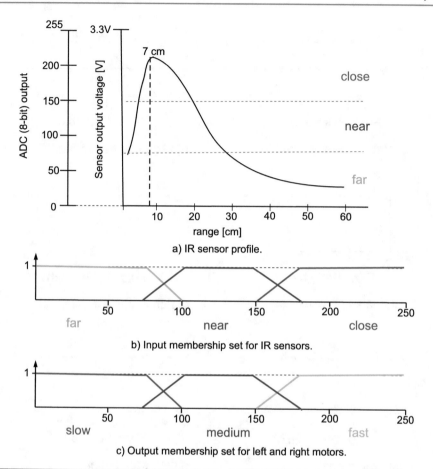

a) IR sensor profile.

b) Input membership set for IR sensors.

c) Output membership set for left and right motors.

d) rule set

e) condensed rule set

Fig. 5.3 Input and output fuzzy membership function development

5.2.3 Apply Rules

A set of rules of the form "IF (antecedent)–THEN (consequent)" are now developed to link input membership function linguistic variables to desired output membership function values. Specific rules are developed by considering different combinations of the input membership function linguistic variables to form the antecedent. Multiple input membership function linguistic variables may be linked using "AND" and "OR" logic connectives. For the "AND" connective, the minimum value of the input membership function linguistic variable is chosen. For the "OR" connective, the maximum value of the input membership function linguistic variable is chosen. The desired output (consequent) for a given combination of input variables is determined by an expert (you–the system designer). The output membership functions are determined by linking the output linguistic variables to desired crisp numerical values.

Example: In Fig. 5.3c, we have developed the output membership functions for the left and right motor. The crisp numerical output values range from 0 to 250. These will serve as inputs to a pulse width modulation (PWM) function to control the left and right motor speed to render different turns.

To construct the rules, linking inputs to outputs, the combination of input linguistic variables for the right and front sensor are placed in a table. The desired output for each combination of inputs is then determined by an expert (you). For example, for the right sensor at "r_close" and the front sensor at "f_close," the desired robot action is a left medium turn ("l_med"). This is accomplished by setting the left motor to slow ("l_slow") and the right motor to medium ("r_med"). The resulting table is provided in Fig. 5.3d.

5.2.4 Aggregate Active Rules and Defuzzify Output

To determine a crisp output(s) to control the system, all of the rules that have provided an output are aggregated together to provide a single output. There are a variety of methods available to do this. The Arduino Embedded Fuzzy Logic Library (eFLL) uses the Mamdani Minimum technique to aggregate the outputs and the center of area technique to defuzzify to crisp outputs [3, 7, 9, 10].

Example: Based on the processes described, two crisp numerical outputs are provided to render the desired motor turn to avoid maze walls.

In the Application section we develop the actual code using the Arduino Embedded Fuzzy Logic Library (eFLL).

5.3   Arduino eFLL

We next explore the Arduino Embedded Fuzzy Logic Library (eFLL). The library was developed by a team (Alves, Lira, Lemos, Kridi, and Leal) of the Robotic Research Group at the State University of Piaui in Tersini, Piaui, Brazil. The library contains features and representative examples to design complex, fuzzy logic control systems. The examples may serve as templates to design other systems. The team has provided an outstanding service of making fuzzy logic concepts readily accessible to the Arduino community (Alves).

We now explore the eFLL Fuzzy Logic Library examples in some detail.

5.3.1   Example: Simple

Early in the chapter we discussed an obstacle avoidance system for vehicle control. Intuitively, we desired a system that would respond differently for an obstruction that is very close rather than much further away. For example, if a large animal (e.g. an elk) steps out right in front of my vehicle, I would like the control system to rapidly apply strong brake pressure to bring the vehicle to a controlled but abrupt stop. On the other hand, if the control system detects an obstacle much farther away from the car, the brake system may be gently applied to slow the car a little bit.

The eFLL provides an example called "Arduino_simple_sample" that implements such a system. The system contains a single input called "input:distance" with linguistic variables "small," "safe," and "big." The input membership function is shown in Fig. 5.4. The system contains a single output called "output:speed" with linguistic variables "slow," "average," and "fast." The output membership function is shown in Fig. 5.4 (Alves).

The system implements the following rules:

- Rule 1: IF distance = small THEN speed = slow
- Rule 2: IF distance = safe THEN speed = average
- Rule 3: IF distance = big THEN speed = high

It is highly recommended the reader upload and run the "Arduino_simple_sample" sketch and become acquainted with its operation. The modified version of this sketch is provided below. In place of a random number generator to generate system input, a potentiometer is used to simulate a distance sensor input and an LED is used as a speed indicator output as shown in Fig. 5.5. Also, additional code steps have been added to report the pertinence (strength) of the fuzzy linguistic variables of the input and output membership functions.

```
//*********************************************************
//arduino_simple_sample_adc: this sketch provides a
//basic example of the Arduino Embedded Fuzzy Logic
//Library (eFLL).
```

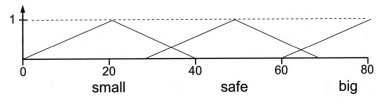

input: distance

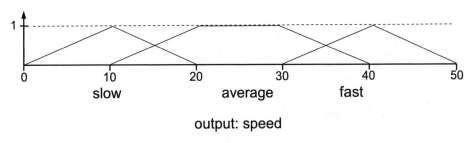

output: speed

Fig. 5.4 Fuzzy inputs and outputs

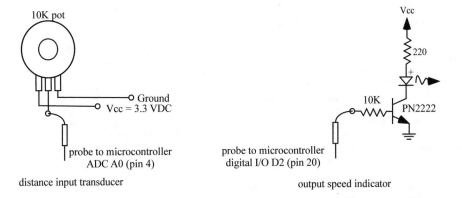

Fig. 5.5 Simulator for input and output

```
//
//This sketch simulates a speed control feature of an
//automatic braking system.
//- Fuzzy Input: distance to obstruction (e.g. another
//   vehicle or an animal)
//- Fuzzy Output: vehicle speed
//
//Author of eFLL Library: A.J. Alves
//
//Example adapted to include sensor speed input,
//output to an LED, and display of input and output
//pertinence
```

```
//- Use 10K pot on ADC A0 (pin 4) as input distance
//- Use LED on D3 (pin 21) to indicate speed
//*******************************************************

#include <Fuzzy.h>

#define speed_LED    3          //physical pin 21

//Instantiating a Fuzzy object
Fuzzy *fuzzy = new Fuzzy();

//Fuzzy Input

//Instantiating a FuzzySet objects
FuzzySet *small = new FuzzySet(0, 20, 20, 40);
FuzzySet *safe = new FuzzySet(30, 50, 50, 70);
FuzzySet *big = new FuzzySet(60, 80, 80, 80);

//Fuzzy Output
//Instantiating a FuzzySet object
FuzzySet *slow = new FuzzySet(0, 10, 10, 20);
FuzzySet *average= new FuzzySet(10, 20, 30, 40);
FuzzySet *fast = new FuzzySet(30, 40, 40, 50);


void setup()
{
Serial.begin(9600);                        //Set the Serial output
//randomSeed(analogRead(0));               //Set a random seed

//Instantiating a FuzzyInput object
//Including the FuzzySet into FuzzyInputs
FuzzyInput *distance = new FuzzyInput(1);
distance->addFuzzySet(small);
distance->addFuzzySet(safe);
distance->addFuzzySet(big);
fuzzy->addFuzzyInput(distance);

//Instantiating a FuzzyOutput objects
//Including the FuzzySet into FuzzyOutputs
FuzzyOutput *speed = new FuzzyOutput(1);
speed->addFuzzySet(slow);
speed->addFuzzySet(average);
speed->addFuzzySet(fast);
fuzzy->addFuzzyOutput(speed);

//Building FuzzyRule "IF distance = small THEN speed = slow"
//Instantiating a FuzzyRuleAntecedent objects
FuzzyRuleAntecedent *ifDistanceSmall = new FuzzyRuleAntecedent();
//Creating a FuzzyRuleAntecedent with just a single FuzzySet
ifDistanceSmall->joinSingle(small);
//Instantiating a FuzzyRuleConsequent objects
FuzzyRuleConsequent *thenSpeedSlow = new FuzzyRuleConsequent();
//Including a FuzzySet to this FuzzyRuleConsequent
thenSpeedSlow->addOutput(slow);
//Instantiating a FuzzyRule objects
```

```
FuzzyRule *fuzzyRule01 = new FuzzyRule(1, ifDistanceSmall, thenSpeedSlow);
//Including the FuzzyRule into Fuzzy
fuzzy->addFuzzyRule(fuzzyRule01);

//Building FuzzyRule "IF distance = safe THEN speed = average"
//Instantiating a FuzzyRuleAntecedent objects
FuzzyRuleAntecedent *ifDistanceSafe = new FuzzyRuleAntecedent();
//Creating a FuzzyRuleAntecedent with just a single FuzzySet
ifDistanceSafe->joinSingle(safe);
//Instantiating a FuzzyRuleConsequent objects
FuzzyRuleConsequent *thenSpeedAverage = new FuzzyRuleConsequent();
//Including a FuzzySet to this FuzzyRuleConsequent
thenSpeedAverage->addOutput(average);
//Instantiating a FuzzyRule objects
FuzzyRule *fuzzyRule02 = new FuzzyRule(2, ifDistanceSafe, thenSpeedAverage);
//Including the FuzzyRule into Fuzzy
fuzzy->addFuzzyRule(fuzzyRule02);

//Building FuzzyRule "IF distance = big THEN speed = high"
//Instantiating a FuzzyRuleAntecedent objects
FuzzyRuleAntecedent *ifDistanceBig = new FuzzyRuleAntecedent();
//Creating a FuzzyRuleAntecedent with just a single FuzzySet
ifDistanceBig->joinSingle(big);
//Instantiating a FuzzyRuleConsequent objects
FuzzyRuleConsequent *thenSpeedFast = new FuzzyRuleConsequent();
//Including a FuzzySet to this FuzzyRuleConsequent
thenSpeedFast->addOutput(fast);
//Instantiating a FuzzyRule objects
FuzzyRule *fuzzyRule03 = new FuzzyRule(3, ifDistanceBig, thenSpeedFast);
//Including the FuzzyRule into Fuzzy
fuzzy->addFuzzyRule(fuzzyRule03);
}

void loop()
{
//Getting a random value
//int input = random(0, 80);

int input_sensor = analogRead(0);
int input = map(input_sensor, 0, 1023, 0, 80);

//Printing something
Serial.println("\n\n\nEntrance: ");
Serial.print("\t\t\tDistance: ");
Serial.println(input);
//Set the random value as an input
fuzzy->setInput(1, input);
//Running the Fuzzification
fuzzy->fuzzify();
//Running the Defuzzification
float output = fuzzy->defuzzify(1);
//Printing something
Serial.println("Result: ");
Serial.print("\t\t\tSpeed: ");
Serial.println(output);
```

```
Serial.println("Input: ");
Serial.print("\tDistance: small-> ");
Serial.print(small->getPertinence());
Serial.print(", safe-> ");
Serial.print(safe->getPertinence());
Serial.print(", big-> ");
Serial.println(big->getPertinence());

Serial.print("\tSpeed: slow-> ");
Serial.print(slow->getPertinence());
Serial.print(",  average-> ");
Serial.print(average->getPertinence());
Serial.print(",  fast-> ");
Serial.print(fast->getPertinence());

//illuminate LED at intensity consistent with speed
int output_int = (int)(output); //cast output speed to int
//analogWrite range 0 to 255
int output_int_PWM = map(output_int, 0, 80, 0, 255);
analogWrite(speed_LED, output_int_PWM);

//wait 2 seconds
delay(2000);
}

//*******************************************************
```

To test the implemented fuzzy logic controller, several test runs are examined as shown in Fig. 5.6. Results are provided for three test runs. It is recommended the reader apply the three control rules with the pertinence (strength) data provided for each input and output membership functions to verify the operation of the controller. Recall, in the defuzzification step the controller applies the Mamdani minimum approach to aggregate the rules that have fired and then performs the center–of–area calculation to determine the crisp output [7].

5.3.2   Example: Advanced

This example provides for a more complex vehicle control system. This example is also provided in the eFLL Library and is called "Arduino_advanced_sample." The system contains three inputs called: "input:distance," "input:speedInput," and "input:temperature." The linguistic variables for each input memebership function are shown in Fig. 5.7. The system also contains two output variables called "output:risk" and "output:speedOutput" with linguistic variables shown in Fig. 5.7.

The system implements the following rules:

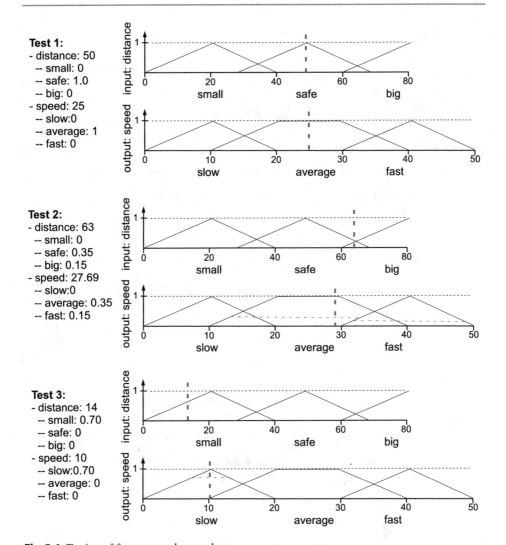

Fig. 5.6 Testing of fuzzy control examples

- Rule 1: ifDistanceNearAndSpeedQuickOrTemperatureCold, thenRiskMaximumAnd SpeedSlow.
- Rule 2: ifDistanceSafeAndSpeedNormalOrTemperatureGood, thenRiskAverageAnd SpeedNormal.
- Rule 3: ifDistanceDistantAndSpeedSlowOrTemperatureHot, thenRiskMinimumSpeed Quick.

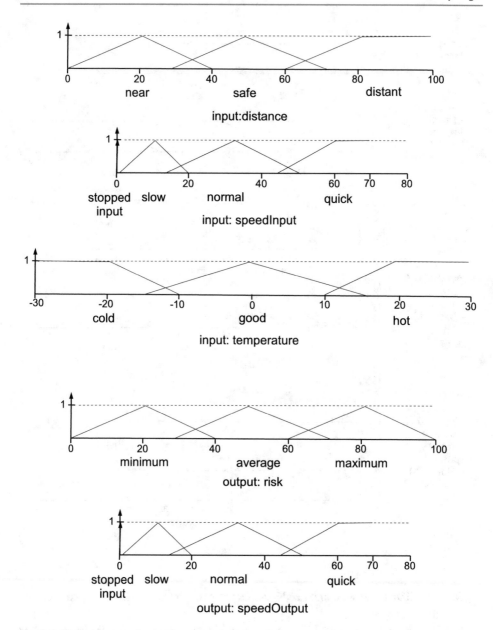

Fig. 5.7 Fuzzy inputs and outputs

It is highly recommended the reader upload and run the "Arduino_advanced_sample" sketch and become acquainted with its operation. In the interest of space, the code will not be duplicated here.

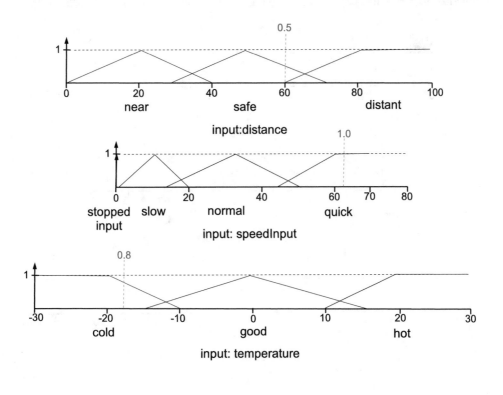

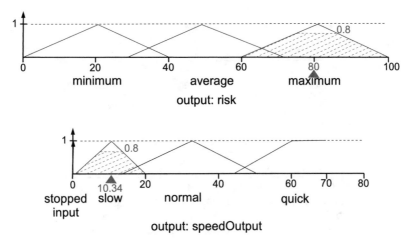

Fig. 5.8 Fuzzy inputs and outputs Test 1

The code provides simulated sensor input using random numbers. The results of several test runs are provided in Figs. 5.8 and 5.9. It is recommended the reader apply the control rules with the pertinence (strength) data provided for each input and output membership functions to verify the operation of the controller. For a complex rule including "AND" and

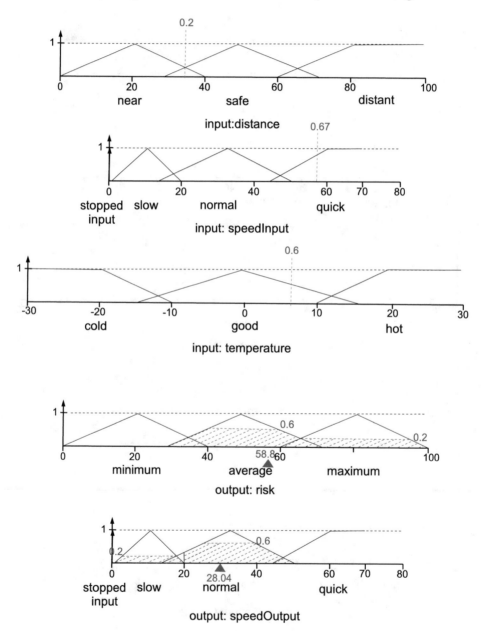

Fig. 5.9 Fuzzy inputs and outputs Test 2

"OR" logical operators, recall the minimum value is chosen for the "AND" operation and the maximum value is chosen for the "OR" Also, as a friendly reminder, in the defuzzification step the controller applies the Mamdani minimum approach to aggregate the rules that have fired and then performs a center–of–area calculation to determine the crisp outputs (Alves).

5.4   Application

In the Application section of Chap. 3, we equipped the Dagu Magician Robot with three IR sensors to detect maze walls. A series of traditional (nonfuzzy) "IF–THEN" statements were used to determine motor action to navigate through the maze without bumping into maze walls.

We revisited the robot earlier in this chapter in our discussion on fuzzy logic theory. We now implement and test a fuzzy logic controller for the robot using the eFLL "Arduino_advanced_sample" as a template guide. We test the robot using simulated IR sensor inputs and motor control outputs as shown in Fig. 5.10. Note we implement the fuzzy control algorithm on the Arduino UNO R3. This processor hosts 32K bytes of Flash memory and 2K bytes of RAM.

In the fuzzy sketch development we limit our sketch development to two (input and right facing) IR sensors. We leave the use of three IR sensors as a homework assignment. Before continuing, please review the development of input measurement functions, output membership functions, and rule development provided earlier in the chapter.

As shown in Fig. 5.3d, the nine rules developed to control the robot include:

- Rule 1: IF (right sensor close) AND (front sensor close) THEN (left motor slow) AND (right motor medium)
- Rule 2: IF (right sensor close) AND (front sensor near) THEN (left motor slow) AND (right motor slow)
- Rule 3: IF (right sensor close) AND (front sensor far) THEN (left motor fast) AND (right motor fast)
- Rule 4: IF (right sensor near) AND (front sensor close) THEN (left motor slow) AND (right motor slow)
- Rule 5: IF (right sensor near) AND (front sensor near) THEN (left motor slow) AND (right motor slow)
- Rule 6: IF (right sensor near) AND (front sensor far) THEN (left motor fast) AND (right motor fast)
- Rule 7: IF (right sensor far) AND (front sensor close) THEN (left motor medium) AND (right motor slow)
- Rule 8: IF (right sensor far) AND (front sensor near) THEN (left motor slow) AND (right motor slow)

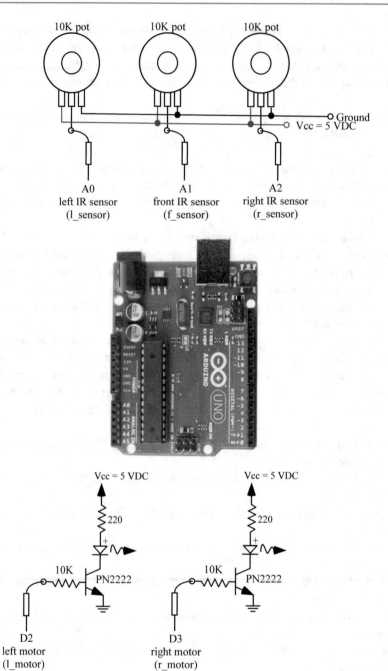

Fig. 5.10 Simulator for inputs and outputs

- Rule 9: IF (right sensor far) AND (front sensor far) THEN (left motor fast) AND (right motor fast)

For proper sketch operation, rules with the same consequent should be combined into a single rule as shown in Fig. 5.3e. This allows the robot to be controlled with four rules. The rules may be simplified further by carefully examining the rule table. For example, when the front sensor is reporting f_far, it does not matter what the value of the right sensor is reporting, since the resulting robot action is the same. This results in a simplified rule set of:

- Rule 1: IF (right sensor close) AND (front sensor close) THEN (left motor slow) AND (right motor medium)
- Rule 2: IF (front sensor near) OR ((right sensor close) AND (front sensor close)) THEN (left motor slow) AND (right motor slow)
- Rule 3: IF (front sensor far) THEN (left motor fast) AND (right motor fast)
- Rule 4: IF (right sensor far) AND (front sensor close) THEN (left motor medium) AND (right motor slow)

The sketch to implement the fuzzy control algorithm is provided next.

```
//****************************************************************
//two_sensor_fuzzy_robot: provides fuzzy control for a two
//sensor robot (front and right facing) for autonomous maze
//navigation.
//
//eFLL example arduino_advanced_sample (Alves) used as
//template for development.
//
//Fuzzy control inputs
//- left  IR sensor   A0 - not used in this sketch
//- front IR sensor   A1
//- right IR sensor   A2
//
//Fuzzy control outputs
//- left  motor   9
//- right motor  10
//****************************************************************

#include <Fuzzy.h>
#include <FuzzyComposition.h>
#include <FuzzyInput.h>
#include <FuzzyIO.h>
#include <FuzzyOutput.h>
#include <FuzzyRule.h>
#include <FuzzyRuleAntecedent.h>
#include <FuzzyRuleConsequent.h>
#include <FuzzySet.h>

//define robot conncetions
#define left_IR_sensor    A0 //pin 4
#define front_IR_sensor   A1 //pin 5
#define right_IR_sensor   A2 //pin 6
```

```
#define left_motor        9
#define right_motor       10

//indicate use of simulator (0) or random generator (1)
bool simulator = 1;
int input1, input2;

//Declare membership functions globally for pertinence access.
//Instantiate all objects in loop()

// Fuzzy
Fuzzy *fuzzy = new Fuzzy();

//FuzzyInput - right IR sensor
FuzzySet *r_far   = new FuzzySet(  0,  0, 75,100);
FuzzySet *r_near  = new FuzzySet( 75,100,150,175);
FuzzySet *r_close = new FuzzySet(150,175,255,255);

//FuzzyInput - front IR sensor
FuzzySet *f_far   = new FuzzySet(  0,  0, 75,100);
FuzzySet *f_near  = new FuzzySet( 75,100,150,175);
FuzzySet *f_close = new FuzzySet(150,175,255,255);

//FuzzyOutput - left motor
FuzzySet *l_slow  = new FuzzySet(  0,  0, 75,100);
FuzzySet *l_medium= new FuzzySet( 75,100,150,175);
FuzzySet *l_fast  = new FuzzySet(150,175,255,255);

//FuzzyOutput - right motor
FuzzySet *r_slow  = new FuzzySet(  0,  0, 75,100);
FuzzySet *r_medium= new FuzzySet( 75,100,150,175);
FuzzySet *r_fast  = new FuzzySet(150,175,255,255);

//sensor readings
//int left_IR_sensor_value;   //left IR sensor variable
int front_IR_sensor_value;    //front IR sensor variable
int right_IR_sensor_value;    //right IR sensor variable

void setup()
{
//Set the Serial output
Serial.begin(9600);

//Set a random seed
randomSeed(analogRead(0));

//FuzzyInput - right IR sensor r_sensor
FuzzyInput *r_sensor = new FuzzyInput(1);
r_sensor->addFuzzySet(r_far);
r_sensor->addFuzzySet(r_near);
r_sensor->addFuzzySet(r_close);
fuzzy->addFuzzyInput(r_sensor);

//FuzzyInput - front IR sensor f_sensor
FuzzyInput *f_sensor = new FuzzyInput(2);
f_sensor->addFuzzySet(f_far);
f_sensor->addFuzzySet(f_near);
f_sensor->addFuzzySet(f_close);
fuzzy->addFuzzyInput(f_sensor);
```

```
//FuzzyOutput - left motor
FuzzyOutput *l_motor = new FuzzyOutput(1);
l_motor->addFuzzySet(l_slow);
l_motor->addFuzzySet(l_medium);
l_motor->addFuzzySet(l_fast);
fuzzy->addFuzzyOutput(l_motor);

//FuzzyOutput - right motor
FuzzyOutput *r_motor = new FuzzyOutput(2);
r_motor->addFuzzySet(r_slow);
r_motor->addFuzzySet(r_medium);
r_motor->addFuzzySet(r_fast);
fuzzy->addFuzzyOutput(r_motor);

//Fuzzy Rule Set
//Building Fuzzy Rule 1
//Antecedent
FuzzyRuleAntecedent *rsensorcloseAndfsensorclose = new FuzzyRuleAntecedent();
rsensorcloseAndfsensorclose->joinWithAND(r_close, f_close);
//Consequent
FuzzyRuleConsequent *thenlmotorslowAndrmotormed = new FuzzyRuleConsequent();
thenlmotorslowAndrmotormed->addOutput(l_slow);
thenlmotorslowAndrmotormed->addOutput(r_medium);
//Assemble Rule
FuzzyRule *fuzzyRule1 = new FuzzyRule(1,rsensorcloseAndfsensorclose,
                                      thenlmotorslowAndrmotormed);
fuzzy->addFuzzyRule(fuzzyRule1);

//Building Fuzzy Rule 2
//Antecedent
FuzzyRuleAntecedent *fsensornear = new FuzzyRuleAntecedent();
fsensornear->joinSingle(f_near);
FuzzyRuleAntecedent *rsensornearAndfsensorclose = new FuzzyRuleAntecedent();
rsensornearAndfsensorclose->joinWithAND(r_near, f_close);
FuzzyRuleAntecedent *ifrsensornearAndfsensorcloseOrfsensornear =
    new FuzzyRuleAntecedent();
ifrsensornearAndfsensorcloseOrfsensornear->
    joinWithOR(rsensornearAndfsensorclose, fsensornear);
//Consequent
FuzzyRuleConsequent *thenlmotorslowAndrmotorslow = new FuzzyRuleConsequent();
thenlmotorslowAndrmotorslow->addOutput(l_slow);
thenlmotorslowAndrmotorslow->addOutput(r_slow);
//Assemble Rule
FuzzyRule *fuzzyRule2 = new FuzzyRule(2,ifrsensornearAndfsensorcloseOrfsensornear,
    thenlmotorslowAndrmotorslow);
fuzzy->addFuzzyRule(fuzzyRule2);

//Building Fuzzy Rule 3
//Antecedent
FuzzyRuleAntecedent *fsensorfar = new FuzzyRuleAntecedent();
fsensorfar->joinSingle(f_far);
//Consequent
FuzzyRuleConsequent *thenlmotorfastAndrmotorfast = new FuzzyRuleConsequent();
thenlmotorfastAndrmotorfast->addOutput(l_fast);
thenlmotorfastAndrmotorfast->addOutput(r_fast);
//Assemble Rule
FuzzyRule *fuzzyRule3 = new FuzzyRule(3,fsensorfar,thenlmotorfastAndrmotorfast);
fuzzy->addFuzzyRule(fuzzyRule3);
```

```
//Building Fuzzy Rule 4
//Antecedent
FuzzyRuleAntecedent *rsensorfarAndfsensorclose = new FuzzyRuleAntecedent();
rsensorfarAndfsensorclose->joinWithAND(r_far, f_close);
//Consequent
FuzzyRuleConsequent *thenlmotormedAndrmotorslow = new FuzzyRuleConsequent();
thenlmotormedAndrmotorslow->addOutput(l_medium);
thenlmotormedAndrmotorslow->addOutput(r_slow);
//Assemble Rule
FuzzyRule *fuzzyRule4 = new FuzzyRule(4,rsensorfarAndfsensorclose,
      thenlmotormedAndrmotorslow);
fuzzy->addFuzzyRule(fuzzyRule4);

//motor pin configuration
pinMode(left_motor, OUTPUT);
pinMode(right_motor, OUTPUT);
}

void loop()
{
if (!simulator)           //use random generator for input
  {
  //get random entrance value for right and front IR sensors
  input1 = random(0, 255);
  input2 = random(0, 255);
  }
else
  {                        //use potentiometer input
  //read analog output from IR sensors
  //left_IR_sensor_value  = analogRead(left_IR_sensor); //pin 4 A0
  front_IR_sensor_value = analogRead(front_IR_sensor);  //pin 5 A1
  right_IR_sensor_value = analogRead(right_IR_sensor);  //pin 6 A2
                          //remap from 1023 to 255 scale
  input1 = map(right_IR_sensor_value, 0, 1023, 0, 255);
  input2 = map(front_IR_sensor_value, 0, 1023, 0, 255);
  }

Serial.print("R sensor:  ");
Serial.print(input1);
Serial.print("\t, F sensor:  ");
Serial.println(input2);
//Serial.println(" ");

fuzzy->setInput(1, input1);
fuzzy->setInput(2, input2);

fuzzy->fuzzify();

//right sensor
//Serial.println("R Input: ");
Serial.print("R sensor: close-> ");
Serial.print(r_close->getPertinence());
Serial.print(", near-> ");
Serial.print(r_near->getPertinence());
Serial.print(", far-> ");
Serial.println(r_far->getPertinence());
//Serial.println(" ");
```

```
//front sensor
//Serial.println("F Input: ");
Serial.print("F sensor: close-> ");
Serial.print(f_close->getPertinence());
Serial.print(", near-> ");
Serial.print(f_near->getPertinence());
Serial.print(", far-> ");
Serial.println(f_far->getPertinence());
//Serial.println(" ");

float output1 = fuzzy->defuzzify(1);      //left motor
float output2 = fuzzy->defuzzify(2);      //right motor

//left motor
//Serial.println("L motor output: ");
Serial.print("L motor: slow-> ");
Serial.print(l_slow->getPertinence());
Serial.print(", Medium-> ");
Serial.print(l_medium->getPertinence());
Serial.print(", Fast-> ");
Serial.println(l_fast->getPertinence());
//Serial.println(" ");

//right motor
//Serial.println("R motor output: ");
Serial.print("R motor: slow-> ");
Serial.print(r_slow->getPertinence());
Serial.print(", Medium-> ");
Serial.print(r_medium->getPertinence());
Serial.print(", Fast-> ");
Serial.println(r_fast->getPertinence());
//Serial.println(" ");

if(simulator)
  {
  //motor control - left motor
  int output1_int = (int)(output1);        //LED on pin 9
  analogWrite(left_motor, output1_int);
  Serial.println("L motor output: ");
  Serial.print(output1_int);
  Serial.println(" ");

  //motor control - right motor
  int output2_int = (int)(output2);        //LED on pin 10
  analogWrite(right_motor, output2_int);
  Serial.println("R motor output: ");
  Serial.print(output2_int);
  Serial.println(" ");
  Serial.println(" ");
  }

//wait 2 seconds
delay(2000);
}

//*************************************************************
```

To test the fuzzy robot controller, use the simulator to test all combinations of sensor inputs provided in the rule table of Fig. 5.3d. Once you have the basic controller operating correctly, consider adding left facing sensor to detect maze walls.

5.5   Summary

This chapter described how to control a process using fuzzy logic techniques. We explored fuzzy logic that allows multiple levels of truth between logic one and zero. We found that many real world control problems lend themselves to fuzzy logic implementation. We then investigated the design of fuzzy logic controllers in some detail. We began with a brief review of the key concepts and design of a fuzzy logic controller. We then explored the Arduino Embedded Fuzzy Logic Library (eFLL) and examples developed by a team of the Robotic Research Group at the State University of Piaui in Tersini, Piaui, Brazil. We explore the examples in some detail. In an earlier writing project, In the Application section, we equipped the Dagu Magician Robot with an eFLL based fuzzy logic control system.

5.6   Problems

1. In your own words compare and contrast the traditional and fuzzy logic approach to controller design.
2. Provide a sketch of the fuzzy controller design process. Briefly describe what activities are required at each step.
3. Illustrate the fuzzy control design process with your own example.
4. How are input membership functions derived from crisp sensor outputs.
5. Describe the different types of trapezoids available to define input and output membership functions.
6. How are output membership defined for a specific application?
7. Describe the use of AND and OR operators in developing rules.
8. What is the basic configuration of a fuzzy logic rule.
9. What are antecedents and consequents? How are they used in fuzzy logic rule development?
10. In your own words describe how the fuzzy outputs are converted to crisp system outputs.
11. In the Application section of this chapter, we provided the design of a two sensor maze following robot. Revise the design to include a third left facing sensor.

References

1. A.J. Alves, R. Lira, M. Lemos, D.S. Kridi, and K. Leal, *Arduino Embedded Fuzzy Logic Library (eFLL)*, Robotic Research Group at the State University of Piaui, Tersini, Piaui, Brazil.
2. A.J. Alves, *eFLL–A Fuzzy Library for Arduino and Embedded Systems*, https://www.blog.zerokol.com.
3. T. Jiang and Y. Li, *Generalized Defuzzification Strategies and Their Parameter Learning Procedures*, IEEE Transactions on Fuzzy Systems, Vol. 4, No. 1, February 1996, 64-71.
4. A.D. Kulkarni, *Computer Vision and Fuzzy–Neural Systems*, Prentice Hall, 2001.
5. C.C. Lee, *Fuzzy Logic in Control Systems: Fuzzy Logic Controller–Part I*, IEEE Transactions on Systems, Man, and Cybernetics, Vol. 20, No. 2, March/April 1990, pp. 404-418.
6. C.C. Lee, *Fuzzy Logic in Control Systems: Fuzzy Logic Controller–Part I*, IEEE Transactions on Systems, Man, and Cybernetics, Vol. 20, No. 2, March/April 1990, pp. 419-435.
7. E.H. Mamdani and S. Assilian, *An Experiment in Linguistic Synthesis with a Fuzzy Logic Controller*, Int. J. Man–Machine Studies, (1975), 1–13.
8. D.J. Pack and S.F. Barrett, *68HC12 Microcontroller Theory and Application*, Prentice Hall, 2002.
9. D.T. Pham and M. Castellani, *Action aggregation and defuzzification in Mamdani–type fuzzy systems*, Proceedings of the Institution of Mechanical Engineers, Part C: Journal of Mechanical Engineering Science, 2002, 747–759.
10. T.A. Runkler, *Selection of Appropriate Defuzzification Methods Using Application Specific Properties*, IEEE Transactions on Fuzzy Systems, Vol. 5, No. 1, February 1997, 72-79.
11. L.A. Zadeh, *Fuzzy Sets*, Information and Control 8, 338-353, 1965.

Neural Networks

6

Objectives: After reading this chapter, the reader should be able to:

- Describe and sketch a biological neuron;
- Define and provide supporting equations for the perceptron model of the neuron;
- Model the operation of a single perceptron;
- Employ the single perceptron model to linearly classify objects into two different categories;
- Model the operation of a multiple perceptron network;
- Employ the multiple perceptron model to linear classify objects into different categories;
- Define and provide supporting equations for a model of a single neuron;
- Employ the single neuron model to assemble a multiple layer artificial neural network;
- Describe the concept of backpropagation;
- Employ a three–layer, feed–forward network with backpropagation to classify objects into different categories;
- List improvements to the artificial neural network to enhance model convergence; and
- Describe advanced software tools for developing and implementing deep neural networks.

6.1 Overview

In this chapter we explore the concept of neurons and neuron models to solve real world challenges. We begin with a brief description of the biological neuron and investigate a model of the neuron, called the perceptron, developed by Frank Rosenblatt in 1959. We use the single perceptron model to separate objects into two categories. We extend the model to include additional perceptrons to separate objects into multiple categories. We then investigate the

© The Author(s), under exclusive license to Springer Nature Switzerland AG 2023   149
S. F. Barrett, *Arduino V: Machine Learning*, Synthesis Lectures on Digital Circuits
& Systems, https://doi.org/10.1007/978-3-031-21877-4_6

single neuron and then explore the concept of backpropagation and develop a three–layer feed–forward network with backpropagation. Along the way we develop Arduino sketches of these different models.

6.2   Biological Neuron

Our brains are composed of many neurons. Neurons work together to help us learn, remember, link complex ideas, complete complex tasks, and so many other things. The goal of scientists and engineers have been to understand the operation and interaction of neurons, model their behavior, and use the models to solve complex challenges on computers. Some of the models require massive computing power well beyond the capability of Arduino microcontrollers. However, there are many tasks that can be readily completed on the Arduino Nano 33 BLE Sense and the lower power Arduino UNO R3. We concentrate on these applications in this chapter.

A diagram of a single biological neuron is provided in Fig. 6.1. The neuron's main processing unit is contained within the cell body or soma. The neuron collects information from nearby neurons via a network of input sensors called dendrites. The input information is collected and processed by the soma. If a specific level of accumulated input information is reached, the neuron fires and transmits an electrical signal down its axon. The axon is wrapped with a myelin sheath to aid in signal conduction. When the electrical signal reaches the axon terminal fibers, a chemical transmitter is released to provide information to other nearby neurons [7].

6.3   Perceptron

In 1959 Frank Rosenblatt developed a model of the single perceptron shown in Fig. 6.2 based on the biological neuron. The mathematical description here is based on the excellent development provided in Kulkarni [4].

The perceptron model provides for inputs (x_i) or features that are multiplied by weights (w_i) specific for each input. The weighted inputs along with a bias (offset) are summed to become the net response and then passed to an activation function.

If the net value exceeds a specified value, the perceptron fires and provides an output value (y); otherwise, the output value is 0. In the example shown, the activation threshold value is 0. If the net value is greater than 0, the perceptron provides a 1 output; otherwise the output is 0. Other activation functions may be used.

A single perceptron may be trained to sort objects into two linear separable categories. What does this mean? If you were to plot the objects on a two–dimensional diagram, you could draw a straight line to separate the two different categories of objects. There are many real world technical challenges where this would be useful. For example, we examine a

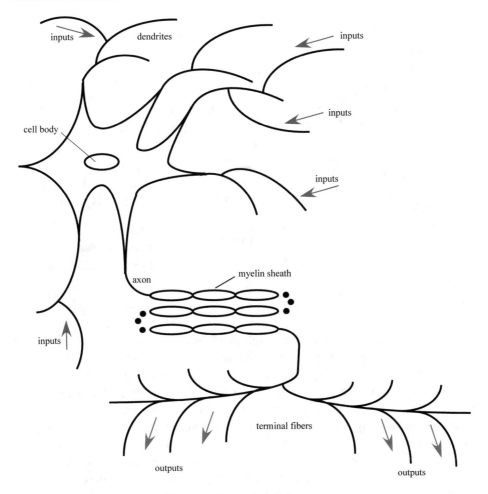

Fig. 6.1 Biological neuron [7]

tomato sorter system in an upcoming example to separate out the largest, most red tomatoes from other varieties. Also, it is important to thoroughly understand perceptron operation as it forms the basis for more complex neuron–based models.

6.3.1 Training the Perceptron Model

To train the perceptron to place objects into two different categories a training set of data is used. The training set contains the relationship between a collection of given input features and the desired output category for those specific inputs. The training set may contain multiple entries of desired input/output pairs.

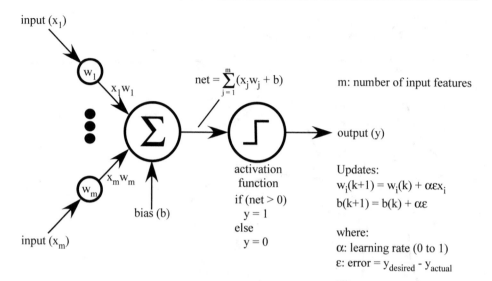

input (x_1)

$$net = \sum_{j=1}^{m}(x_j w_j + b)$$

m: number of input features

output (y)

activation
function
if (net > 0)
  y = 1
else
  y = 0

Updates:
$w_i(k+1) = w_i(k) + \alpha \varepsilon x_i$
$b(k+1) = b(k) + \alpha \varepsilon$

where:
α: learning rate (0 to 1)
ε: error $= y_{desired} - y_{actual}$

bias (b)

input (x_m)

Fig. 6.2 Single perceptron model [4]

To initiate the training sequence, the weights of the perceptron model and the bias are initially set to zero or some small random values. The inputs are then multiplied by the input weights. The weighted inputs are summed with the bias to determine the value of net and then passed to the activation function. The activation function generates the appropriate output (1 or 0) if the net value has exceeded the set threshold of zero.

The perceptron output is compared to the desired output provided by the training set. The error is then calculated. The error is the difference between the desired training set output and the actual output provided by the perceptron output. The weights and the bias are then updated using the equations provided in Fig. 6.2. In addition to the error term, the weights and update equations also provide a learning rate term (alpha). The value of alpha is set to a value from 0 and 1. A larger value provides for more dramatic weight changes. A smaller value of alpha may require additional computation time but potentially yield better convergence results.

The perceptron now processes the second input/output pair from the training set using the updated weights and bias values. This process continues through the entire training set, called the epoch, until the model converges. The model converges when the error for each entry in the training set is zero or at a desired error goal. This may require multiple iterations of applying the training set to the perceptron model. In the first upcoming example, convergence required two sequential applications of the training set. The second example required 1,500 iterative applications of the training set.

Once the model has achieved the error goal for the entire training set, the weight and bias values may be used to plot the line separating the two categories of objects. The overall training process is illustrated in Fig. 6.3. Once the model has been trained, it may be used to categorize new inputs not in the original training set.

Example: In this first perceptron example, we reconstruct the results provided by Dan in "Single Layer Perceptron Explained" in the ML Corner [3]. The perceptron consists of two inputs, a bias, and a single output. The weights and bias are initially set to zero. After two epochs, the mean squared error (MSE) of the output error is zero. The resulting weights and bias may be used to separate the input/output pairs into two different categories as shown in Fig. 6.4.

The line separating category 1 and category 2 objects may be plotted by forming a line equation of the form:

$$x1w1 + x2w2 + b = 0$$

$$x1(-0.1) + x2(0.24) + (-0.1) = 0$$

$$x2 = (0.1)x1/(0.24) + (0.1)/0.24$$

$$y = mx + b$$

$$y = 0.42x + 0.42$$

The following Arduino sketch accomplishes the training task. The UML activity diagram for the sketch is provided in Fig. 6.5. For this example the sketch is run on the Arduino UNO R3. The resulting Arduino sketch output is provided in Fig. 6.6. With the perceptron trained, it may be used to categorize inputs outside the original data set. We will call this the "Run Mode." We discuss this mode in the next section.

```
//****************************************************************
//perceptron_2in_1out_train_run
//Sketch has two modes of operation selected via a DIP switch.
//Sketch should be run first in Train Mode(1) before Run Mode(0).
//attached to input D2:
// - D2 = HIGH, selects training mode
// - D2 = LOW, selects run mode
//
//Other pins used:
// - D3: LED to indicate mode Train(1)/Run(0)
// - D4: LED to indicate inputs placed in Category 1
// - D5: LED to indicate inputs placed in Category 2
//****************************************************************

#include<EEPROM.h>

//define
#define trng_set_size        4     //entries in training set
#define train_run_sw         2     //switch input train(1)/run(0)
```

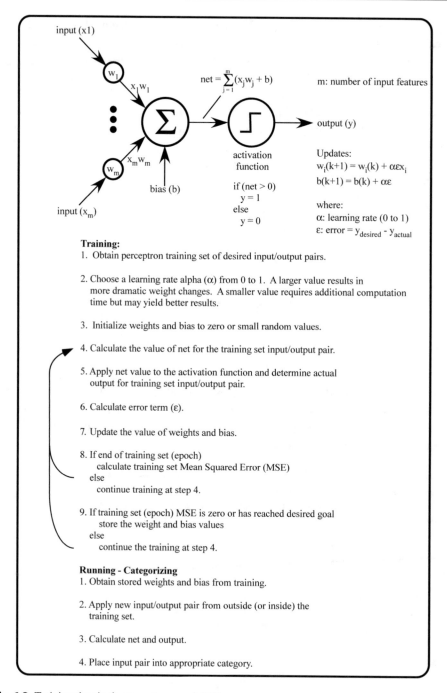

input (x1)

$x_1 w_1$

w_1

$net = \sum_{j=1}^{m}(x_j w_j + b)$

m: number of input features

Σ

$x_m w_m$

w_m

input (x_m)

bias (b)

output (y)

activation
function

if (net > 0)
    y = 1
else
    y = 0

Updates:
$w_i(k+1) = w_i(k) + \alpha \varepsilon x_i$
$b(k+1) = b(k) + \alpha \varepsilon$

where:
α: learning rate (0 to 1)
ε: error $= y_{desired} - y_{actual}$

Training:
1. Obtain perceptron training set of desired input/output pairs.

2. Choose a learning rate alpha (α) from 0 to 1. A larger value results in
   more dramatic weight changes. A smaller value requires additional computation
   time but may yield better results.

3. Initialize weights and bias to zero or small random values.

4. Calculate the value of net for the training set input/output pair.

5. Apply net value to the activation function and determine actual
   output for training set input/output pair.

6. Calculate error term (ε).

7. Update the value of weights and bias.

8. If end of training set (epoch)
       calculate training set Mean Squared Error (MSE)
   else
       continue training at step 4.

9. If training set (epoch) MSE is zero or has reached desired goal
       store the weight and bias values
   else
       continue the training at step 4.

Running - Categorizing
1. Obtain stored weights and bias from training.

2. Apply new input/output pair from outside (or inside) the
   training set.

3. Calculate net and output.

4. Place input pair into appropriate category.

Fig. 6.3 Training the single perceptron model [4]

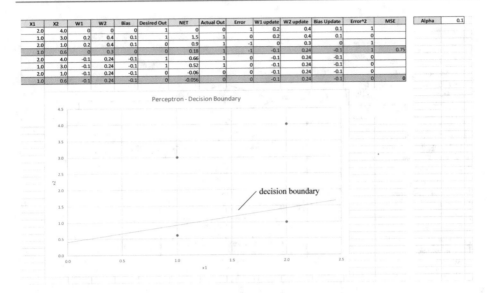

X1	X2	W1	W2	Bias	Desired Out	NET	Actual Out	Error	W1 update	W2 update	Bias Update	Error^2	MSE		Alpha	0.1
2.0	4.0	0	0	0	1	0	0	1	0.2	0.4	0.1	1				
1.0	3.0	0.2	0.4	0.1	1	1.5	1	0	0.2	0.4	0.1	0				
2.0	1.0	0.2	0.4	0.1	1	0.9	1	-1	0	0.3	0	1				
1.0	0.6	0	0.3	0	0	0.18	1	-1	-0.1	0.24	-0.1	1	0.75			
2.0	4.0	-0.1	0.24	-0.1	1	0.66	1	0	-0.1	0.24	-0.1	0				
1.0	3.0	-0.1	0.24	-0.1	1	0.52	1	0	-0.1	0.24	-0.1	0				
2.0	1.0	-0.1	0.24	-0.1	0	-0.06	0	0	-0.1	0.24	-0.1	0				
1.0	0.6	-0.1	0.24	-0.1	0	-0.056	0	0	-0.1	0.24	-0.1	0	0			

Fig. 6.4 Training the single perceptron model Excel spreadsheet [3]

```
#define train_run_LED      3    //LED for train(1)/run(0)
#define cat1_LED           4    //LED for category 1
#define cat2_LED           5    //LED for category 2

//global
float input_1[trng_set_size]   = {2.0, 1.0, 2.0, 1.0};
float input_2[trng_set_size]   = {4.0, 3.0, 1.0, 0.6};
int   desired_output[trng_set_size]   = {1, 1, 0, 0};
int   actual_output;
unsigned int epoch_iterations = 10000;
unsigned int epoch_count = 0;
float w1    = 0;               //set initial weight of w1
float w2    = 0;               //set initial weight of w2
float bias = 0;                //set inital bias value
float learning_rate = 0.1;     //set 0 to 1
float x1, x2;
int   actual_out;              //output from perceptron
float error;                   //desired - actual output
int   train_run_sw_value;      //switch input train(1)/run(0)
float net;                     //numerical out from perceptron
float sq_error = 0;            //squared error
float MSE = 1;                 //mean squared error
int   eeAddress = 0;

void setup()
{
Serial.begin(9600);            //Set the Serial output
                               //set pin modes
pinMode(train_run_sw,    INPUT);
```

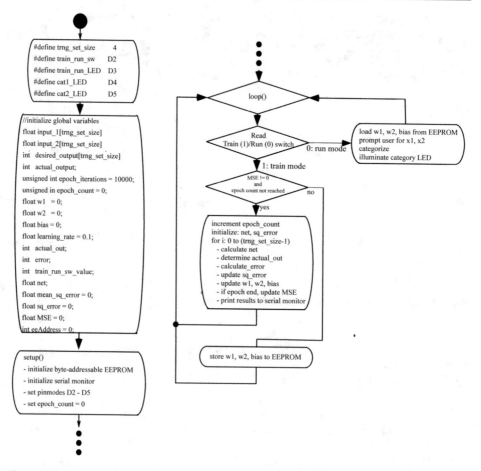

Fig. 6.5 Perceptron train/run sketch UML

```
pinMode(train_run_LED, OUTPUT);
pinMode(cat1_LED, OUTPUT);
pinMode(cat2_LED, OUTPUT);

epoch_count = 0;                        //reset epoch count
}

void loop()
{
int i;

//Reset LEDs
digitalWrite(train_run_LED, LOW);
digitalWrite(cat1_LED, LOW);
digitalWrite(cat2_LED, LOW);
```

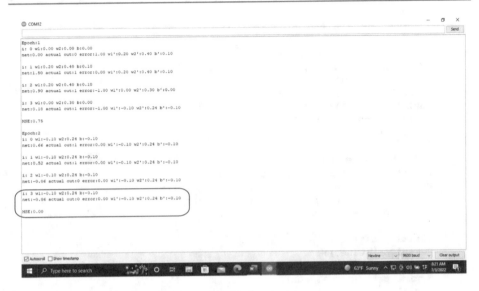

Fig. 6.6 Training the single perceptron model sketch output

```
//Read Train(1)/Run Switch (0)
//Serial.print("Read Train(1)/Run Switch(0):");
train_run_sw_value = digitalRead(train_run_sw);

if(train_run_sw_value == 1)            //training mode
  {
  digitalWrite(train_run_LED, HIGH);
                              //calculate response until MSE = 0
                              //or max count reached
  if((MSE > 0)&&(epoch_count <=epoch_iterations))
    {
    epoch_count = epoch_count + 1;    //epoch count
    Serial.print("Epoch:");
    Serial.println(epoch_count);

    net = 0;                          //reset net
    sq_error = 0;                     //reset sq error
    for(i=0; i < trng_set_size; i++)
      {
      Serial.print("i: ");
      Serial.print(i);
      net = (input_1[i]*w1) + (input_2[i]*w2) + bias;
      if(net > 0)
        {
        actual_out = 1;
        }
      else
        {
```

```
      actual_out = 0;
      }

   Serial.print(" w1:");
   Serial.print(w1);
   Serial.print(" w2:");
   Serial.print(w2);
   Serial.print(" b:");
   Serial.println(bias);

   error = desired_output[i] - actual_out;
   sq_error = sq_error + (error * error);
   w1 = w1 + (learning_rate * error * input_1[i]);
   w2 = w2 + (learning_rate * error * input_2[i]);
   bias = bias + (learning_rate * error);

   Serial.print("net:");
   Serial.print(net);
   Serial.print(" actual out:");
   Serial.print(actual_out);
   Serial.print(" error:");
   Serial.print(error);
   Serial.print(" w1':");
   Serial.print(w1);
   Serial.print(" w2':");
   Serial.print(w2);
   Serial.print(" b':");
   Serial.print(bias);
   Serial.println(" ");

   if(i == (trng_set_size-1))        //end of Epoch
      {
      MSE = sq_error/trng_set_size;
      Serial.println(" ");
      Serial.print("MSE:");
      Serial.print(MSE);
      Serial.println(" ");
      }

   Serial.println(" ");
   }//end for
}//if (MSE)

if(MSE == 0)                          //store weights and bias to EEPROM
   {
   eeAddress = 0;
   EEPROM.put(eeAddress, w1);    //store weight 1
   eeAddress = eeAddress + sizeof(float);
   EEPROM.put(eeAddress, w2);    //store weight 2
   eeAddress = eeAddress + sizeof(float);
   EEPROM.put(eeAddress, bias);  //store bias
   }
}
```

```
else                                     //run mode
  {
  Serial.println("else...");
  digitalWrite(train_run_LED, LOW);

  //get values from EEPROM
  eeAddress = 0;
  EEPROM.get(eeAddress, w1);     //store weight 1
  eeAddress = eeAddress + sizeof(float);
  EEPROM.get(eeAddress, w2);     //store weight 2
  eeAddress = eeAddress + sizeof(float);
  EEPROM.get(eeAddress, bias);  //store bias

  //print results
  Serial.println(" ");
  Serial.print(" w1':");
  Serial.print(w1);
  Serial.print(" w2':");
  Serial.print(w2);
  Serial.print(" b':");
  Serial.print(bias);
  Serial.println(" ");

  //flush input buffer
  while(Serial.available() >0)
    {
    Serial.read();
    }
  //request x1 input value from user via serial monitor
  Serial.println("Insert new value of x1: [send]");
  while(Serial.available()==0){}        //wait for user input data
  x1 = Serial.parseFloat();
  Serial.println(" ");
  Serial.print("x1:");
  Serial.println(x1);

  //flush input buffer
  while(Serial.available() >0)
    {
    Serial.read();
    }
  //request x input value from user via serial monitor
  Serial.print("Insert new value of x2: [send]");
  delay(5000);
  while(Serial.available()==0){}        //wait for user input data
  x2 = Serial.parseFloat();
  Serial.println(" ");
  Serial.print("x2:");
  Serial.println(x2);

                                  //reset LEDs
  digitalWrite(cat1_LED, LOW);
  digitalWrite(cat2_LED, LOW);
```

```
//process new input and assign to appropriate category
//illuminate appropriate LED
net = (x1*w1) + (x2*w2) + bias;
    if(net > 0)
       {
       //category 1
       Serial.println("Category 1");
       digitalWrite(cat1_LED, HIGH);
       digitalWrite(cat2_LED, LOW);
       }
    else
       {
       //category 2
       Serial.println("Category 2");
       digitalWrite(cat1_LED, LOW);
       digitalWrite(cat2_LED, HIGH);
       }
  delay(5000);
  }
}

//*************************************************************
```

6.3.2   Single Perceptron Run Mode

In the previous sketch, the "Run Mode begins at the "else" statement. To implement the "Run Mode," the following additions are required for the sketch:

- Equip the Arduino UNO R3 with an external switch to select between the Train(1)/Run(0) mode.
- Provide an external LED to indicate the current sketch mode: Train(on)/Run(off).
- Provide the sketch the ability to write the value of weights and bias to EEPROM in the Train mode.
- Provide the sketch the ability to read the value of weights and bias from the EEPROM in the Run mode.
- In the Run Mode, have the sketch prompt the user for a new input data pair outside (or inside) the original training set.
- Provide code to categorize the provided data input/output pair into a specific category based on the training set weights and bias.
- Provide external LEDs to indicate the two different object categories.

A UML diagram for the sketch was provided earlier in Fig. 6.5.

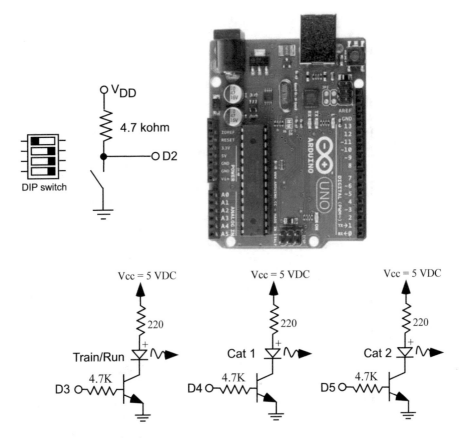

Fig. 6.7 Perceptron test circuit

We use a dual inline package (DIP) switch to select the sketch mode (Train(1)/Run(0)). Also, three LEDs are used to indicate the sketch mode (Train(on)/Run(off)) and the category the new input pair is placed as shown in Fig. 6.7.

To test the sketch, begin by setting the "Train/Run Mode" DIP switch to the "Train" position. When the sketch has completed training (MSE = 0 or desired MSE goal), change the switch position to "Run." The user is prompted for values of x1 and x2 input feature values. To test the perceptron, use input values both inside and outside the original training set. Verify the algorithm places the data into the proper category.

6.3.3 Sorting Tomatoes

Tomatoes are available in a wide variety of color and size. Tomatoes are sorted by their redness and size. These are called input features. Tomatoes that are very red and large

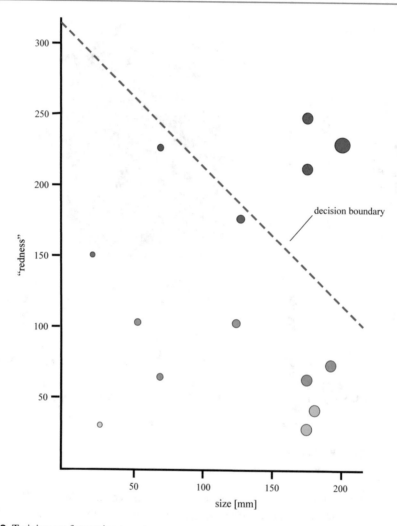

Fig. 6.8 Training set for sorting tomatoes

in diameter garner the best prices. In this example, we develop and train a perceptron to separate the reddest, largest tomatoes from the others. The data set for training the perceptron is provided in Fig. 6.8. We use two different features to categorize tomatoes: redness (0 to 255) and diameter (up to 200 mm).

We use the same process from the previous example to train the perceptron model to obtain values for weights and bias. The sketch is slightly modified to only print results to the serial monitor at the completion of each epoch. The sketch provides the Training Mode. With the perceptron trained, it may be used to categorize inputs outside the original data set. We call this the "Run Mode."

```
//*************************************************************
//tomato_sorter
//Sketch has two modes of operation selected via a DIP switch
//attached to input D0:
// - D0 = HIGH, selects training mode
// - D0 = LOW, selects run mode
//
//Other pins used:
// - D1: LED to indicate mode Train(1)/Run(0)
// - D2: LED to indicate inputs placed in Category 1
// - D3: LED to indicate inputs placed in Category 2
//*************************************************************


//define
#define trng_set_size      12     //entries in training set
#define train_run_sw        0     //switch input train(1)/run(0)
#define train_run_LED       1     //LED for train(1)/run(0)
#define cat1_LED            2     //LED for category 1
#define cat2_LED            3     //LED for category 2
#define print_modulo      100     //print iteration
//global
float input_1[trng_set_size]  = {250, 225, 175, 250, 220, 225, 100, 100,
                                 75, 25, 60, 35};
float input_2[trng_set_size]  = {25, 75, 125, 170, 170, 200, 50, 125,
                                 190, 25, 75, 175};
int   desired_output[trng_set_size]  = {0, 0, 0, 1, 1, 1,
                                        0, 0, 0, 0, 0, 0};
int   actual_output;
unsigned int epoch_iterations = 10000;
unsigned int epoch_count = 0;

float w1   = 0;              //set initial weight of w1
float w2   = 0;              //set initial weight of w2
float bias = 0;              //set initial bias value
float learning_rate = 0.1;   //set 0 to 1

int   actual_out;           //output from perceptron
float error;                //desired - actual output
int   train_run_sw_value;   //switch input train(1)/run(0)
float net;                  //numerical out from perceptron
float sq_error = 0;         //squared error
float MSE = 1;              //mean squared error

void setup()
{
//Initialize byte-addressable EEPROM

//Set the Serial output
Serial.begin(9600);

pinMode(train_run_sw,    INPUT);
pinMode(train_run_LED, OUTPUT);
```

```
pinMode(cat1_LED, OUTPUT);
pinMode(cat2_LED, OUTPUT);

epoch_count = 0;
}

void loop()
{
int i;

//Read Train (1) /Run Switch (0)
//Serial.print("Read Train(1)/Run Switch (0):");
//train_run_sw_value = digitalRead(train_run_sw);
train_run_sw_value = HIGH;
//Serial.println(train_run_sw_value);

if(train_run_sw_value == 1)          //training mode
  {
  if((MSE > 0)&&(epoch_count <=epoch_iterations))
    {
    epoch_count = epoch_count + 1;    //epoch count
    net = 0;                          //reset net
    sq_error = 0;                     //reset sq error
    for(i=0; i < trng_set_size; i++)
      {
      net = (input_1[i]*w1) + (input_2[i]*w2) + bias;
      if(net > 0)
        {
        actual_out = 1;
        }
      else
        {
        actual_out = 0;
        }

      if((i==(trng_set_size-1))&&(epoch_count%print_modulo==0))
        {
        Serial.println(" ");
        Serial.print("Epoch:");
        Serial.println(epoch_count);

        Serial.print("i: ");
        Serial.print(i);

        Serial.print(" w1:");
        Serial.print(w1);

        Serial.print(" w2:");
        Serial.print(w2);

        Serial.print(" b:");
        Serial.println(bias);
        }
```

```
      error = desired_output[i] - actual_out;
      sq_error = sq_error + (error * error);
      w1 = w1 + (learning_rate * error * input_1[i]);
      w2 = w2 + (learning_rate * error * input_2[i]);
      bias = bias + (learning_rate * error);

      if((i==(trng_set_size-1))&&(epoch_count%print_modulo==0))
        {
        Serial.print("net:");
        Serial.print(net);

        Serial.print(" actual out:");
        Serial.print(actual_out);

        Serial.print(" error:");
        Serial.print(error);

        Serial.print(" w1':");
        Serial.print(w1);

        Serial.print(" w2':");
        Serial.print(w2);

        Serial.print(" b':");
        Serial.print(bias);
        }

      if((i==(trng_set_size-1))&&(epoch_count%print_modulo==0))
      //calculate response until MSE = 0
        {
        Serial.println(" ");
        MSE = sq_error/trng_set_size;
        Serial.print("MSE:");
        Serial.print(MSE);
        Serial.println(" ");
        }

      }//end for
   }//if (MSE)

 //store results to EEPROM


  }
else                              //Run Mode
  {
  Serial.println("else...");
  //get values from EEPROM

  //request input values from user via serial monitor

  //process new input
```

```
//illuminate appropriate LED
  }
}
```

```
//************************************************************
```

The training set was applied sequentially to the model 1,500 times to reach convergence. The results of model training are provided in Fig. 6.9.

As before, the values of weights and bias are used to separate categories. The line separating category 1 and category 2 objects may be plotted by forming a line equation of the form:

$$x1w1 + x2w2 + b = 0$$

$$x1(0.5) + x2(0.5) + (-162.70) = 0$$

$$y = mx + b$$

$$y = -x + 325.4$$

The resulting separation line is shown in Fig. 6.8.

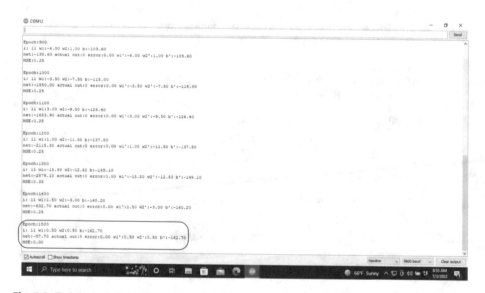

Fig. 6.9 Training the single perceptron model sketch output–tomato sorting

With the perceptron trained, it may be used to categorize inputs outside the original data set. We call this the "Run Mode." As a follow on exercise, modify the tomato sorting sketch to include the "Run" mode.

6.4   Multiple Perceptron Model

The single perceptron model may be used to linearly separate objects into two different categories. Multiple perceptrons may be used to provide for additional categories. A multiple perceptron network is shown in Fig. 6.10. It is important to note the three perceptrons operate independently of one another. That is, they do not share information between them. The perceptrons operate independently but yet in parallel. The perceptrons may be used to place inputs into separate categories as long as the categories are linearly separable.

Provided in Fig. 6.11 is a data set of input/output pairs. The data set is categorized into one of three outputs (Y1, Y2, or Y3). The three perceptron model is trained by applying multiple iterations of the data set to the perceptron network. As shown in Fig. 6.11 over 900 epochs were required for all three perceptrons to converge to a mean squared error of zero using the sketch provided below. Also, the three perceptrons required a different number of epochs to converge.

At the completion of the training portion of the sketch the required weights and biases are provided for each perceptron. As shown in the previous example, the weights and biases are used to form the line equations separating a given category of input/output pairs from those outside the data category. The resulting line equations are provided below and plotted

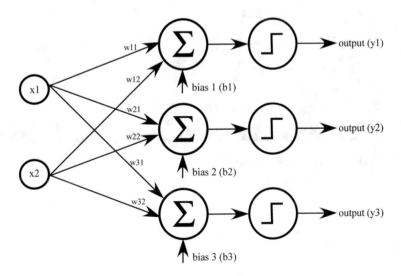

Fig. 6.10 Multiple perceptron model [4]

X1	X2	Y1	Y2	Y3
25	62	1	0	0
12	75	1	0	0
20	80	1	0	0
1	79	1	0	0
80	87	0	1	0
87	80	0	1	0
95	95	0	1	0
76	94	0	1	0
40	17	0	0	1
57	20	0	0	1
35	15	0	0	1
26	4	0	0	1

Y1 X1 < 50, X2 > 50

Y2 X1 > 50, X2 > 50

Y3 ((X1 > 25)&&(X1<75)), X2 < 25

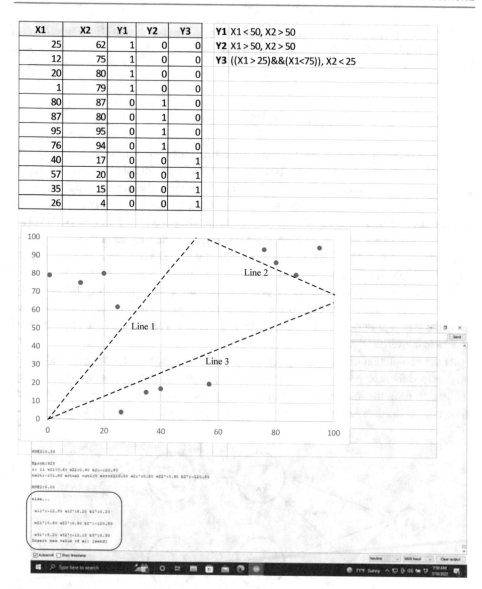

Fig. 6.11 Multiple perceptron model results [4]

in Fig. 6.11. Note how each line separates a given cluster of input/output pairs from the other clusters.

Line 1:

$$x1w1 + x2w2 + b = 0$$

$$x1(-12.8) + x2(6.2) + (0.20) = 0$$

$$y = mx + b$$

$$y = 2.06x + -0.03$$

Line 2:

$$x1w1 + x2w2 + b = 0$$

$$x1(0.60) + x2(0.90) + (-120.8) = 0$$

$$y = mx + b$$

$$y = -0.67x + 134.2$$

Line 3:

$$x1w1 + x2w2 + b = 0$$

$$x1(8.20) + x2(-12.1) + (0.30) = 0$$

$$y = mx + b$$

$$y = 0.67x + 0.02$$

With the perceptron trained, it may be used to categorize inputs outside the original data set. We call this the "Run Mode." We explore this mode in the next section of the chapter.

```
//*****************************************************************
//3perceptron_2in_1out_train_run_final
//Sketch implements 3 perceptron model
//Sketch has two modes of operation: Train and Run
//Model first performs training and then proceeds to run.
//*****************************************************************

//define
#define trng_set_size        12      //entries in training set
```

```
//global
float input_1[trng_set_size]   = {25.0, 12.0, 20.0,  1.0, 80.0, 87.0,
                                  95.0, 76.0, 40.0, 57.0, 35.0, 26.0};
float input_2[trng_set_size]   = {62.0, 75.0, 80.0, 79.0, 87.0, 80.0,
                                  95.0, 94.0, 17.0, 20.0, 15.0,  4.0};
int   desired_output1[trng_set_size]   = {1, 1, 1, 1, 0, 0,
                                          0, 0, 0, 0, 0, 0};
int   desired_output2[trng_set_size]   = {0, 0, 0, 0, 1, 1,
                                          1, 1, 0, 0, 0, 0};
int   desired_output3[trng_set_size]   = {0, 0, 0, 0, 0, 0,
                                          0, 0, 1, 1, 1, 1};
int   actual_output1, actual_output2, actual_output3;
unsigned int epoch_iterations = 15000;
unsigned int epoch_count = 0;
float w11   = 0;              //set init weight of perceptron 1, x1
float w12   = 0;              //set init weight of perceptron 1, x2
float w21   = 0;              //set init weight of perceptron 2, x1
float w22   = 0;              //set init weight of perceptron 2, x2
float w31   = 0;              //set init weight of perceptron 3, x1
float w32   = 0;              //set init weight of perceptron 3, x2
float bias1 = 0, bias2 = 0, bias3 = 0;          //set init bias values
float learning_rate = 0.1;    //set 0 to 1
float x1, x2;
int   actual_out1, actual_out2, actual_out3;  //output from perceptron
float error1, error2, error3;                 //desired - actual output
float net1, net2, net3;                 //numerical out from perceptron
float sq_error1 = 0, sq_error2 = 0, sq_error3 =0;      //squared error
float MSE1 = 1, MSE2 = 1, MSE3 = 1;            //mean squared error

void setup()
{
Serial.begin(9600);          //Set the Serial output
epoch_count = 0;             //reset epoch count
}


void loop()
{
int i;


//Training Mode
if((epoch_count <= epoch_iterations)&&((MSE1 !=0)||(MSE2 !=0)||(MSE3 !=0)))
  {
                                  //calculate response until MSE = 0
                                  //or max count reached
  epoch_count = epoch_count + 1;   //epoch count
  while(!Serial);
  Serial.print("Epoch:");
  Serial.println(epoch_count);
  if((MSE1 > 0)&&(epoch_count <=epoch_iterations))
    {
    net1 = 0;                       //reset net
```

```
sq_error1 = 0;                            //reset sq error
for(i=0; i < trng_set_size; i++)
  {
  if(i == (trng_set_size-1))       //end of Epoch
    {
    Serial.print("i: ");
    Serial.print(i);
    }
  net1 = (input_1[i]*w11) + (input_2[i]*w12) + bias1;
  if(net1 > 0)
    {
    actual_out1 = 1;
    }
  else
    {
    actual_out1 = 0;
    }
  if(i == (trng_set_size-1))       //end of Epoch
    {
    Serial.print(" w11:");
    Serial.print(w11);
    Serial.print(" w12:");
    Serial.print(w12);
    Serial.print(" b1:");
    Serial.println(bias1);
    }

  error1 = desired_output1[i] - actual_out1;
  sq_error1 = sq_error1 + (error1 * error1);
  w11 = w11 + (learning_rate * error1 * input_1[i]);
  w12 = w12 + (learning_rate * error1 * input_2[i]);
  bias1 = bias1 + (learning_rate * error1);

  if(i == (trng_set_size-1))       //end of Epoch
    {
    Serial.print("net1:");
    Serial.print(net1);
    Serial.print(" actual ou1t:");
    Serial.print(actual_out1);
    Serial.print(" error1:");
    Serial.print(error1);
    Serial.print(" w11':");
    Serial.print(w11);
    Serial.print(" w12':");
    Serial.print(w12);
    Serial.print(" b1':");
    Serial.print(bias1);
    Serial.println(" ");
    }

  if(i == (trng_set_size-1))       //end of Epoch
    {
    MSE1 = sq_error1/trng_set_size;
```

```
      Serial.println(" ");
      Serial.print("MSE1:");
      Serial.print(MSE1);
      Serial.println(" ");
      Serial.println(" ");
      }
   }//end for
  }//if (MSE1)

 if((MSE2 > 0)&&(epoch_count <=epoch_iterations))
   {
   net2 = 0;                          //reset net
   sq_error2 = 0;                     //reset sq error
   for(i=0; i < trng_set_size; i++)
     {
     if(i == (trng_set_size-1))       //end of Epoch
       {
       Serial.print("i: ");
       Serial.print(i);
       }
     net2 = (input_1[i]*w21) + (input_2[i]*w22) + bias2;
     if(net2 > 0)
       {
       actual_out2 = 1;
       }
     else
       {
       actual_out2 = 0;
       }

     if(i == (trng_set_size-1))       //end of Epoch
       {
       Serial.print(" w21:");
       Serial.print(w21);
       Serial.print(" w22:");
       Serial.print(w22);
       Serial.print(" b2:");
       Serial.println(bias2);
       }

     error2 = desired_output2[i] - actual_out2;
     sq_error2 = sq_error2 + (error2 * error2);
     w21 = w21 + (learning_rate * error2 * input_1[i]);
     w22 = w22 + (learning_rate * error2 * input_2[i]);
     bias2 = bias2 + (learning_rate * error2);

     if(i == (trng_set_size-1))       //end of Epoch
       {
       Serial.print("ne2t:");
       Serial.print(net2);
       Serial.print(" actual out2:");
       Serial.print(actual_out2);
       Serial.print(" error2:");
```

```
      Serial.print(error2);
      Serial.print(" w21':");
      Serial.print(w21);
      Serial.print(" w22':");
      Serial.print(w22);
      Serial.print(" b2':");
      Serial.print(bias2);
      Serial.println(" ");
      }

   if(i == (trng_set_size-1))      //end of Epoch
      {
      MSE2 = sq_error2/trng_set_size;
      Serial.println(" ");
      Serial.print("MSE2:");
      Serial.print(MSE2);
      Serial.println(" ");
      Serial.println(" ");
     }
   }//end for
  }//if (MSE2)

 if((MSE3 > 0)&&(epoch_count <=epoch_iterations))
   {
   net3 = 0;                             //reset net
   sq_error3 = 0;                        //reset sq error
   for(i=0; i < trng_set_size; i++)
     {
     if(i == (trng_set_size-1))      //end of Epoch
        {
        Serial.print("i: ");
        Serial.print(i);
        }
      net3 = (input_1[i]*w31) + (input_2[i]*w32) + bias3;
      if(net3 > 0)
         {
         actual_out3 = 1;
         }
      else
         {
         actual_out3 = 0;
         }
      if(i == (trng_set_size-1))      //end of Epoch
         {
         Serial.print(" w31:");
         Serial.print(w31);
         Serial.print(" w32:");
         Serial.print(w32);
         Serial.print(" b3:");
         Serial.println(bias3);
         }

      error3 = desired_output3[i] - actual_out3;
```

```
      sq_error3 = sq_error3 + (error3 * error3);
      w31 = w31 + (learning_rate * error3 * input_1[i]);
      w32 = w32 + (learning_rate * error3 * input_2[i]);
      bias3 = bias3 + (learning_rate * error3);

      if(i == (trng_set_size-1))        //end of Epoch
        {
        Serial.print("net3:");
        Serial.print(net3);
        Serial.print(" actual out3:");
        Serial.print(actual_out3);
        Serial.print(" error3:");
        Serial.print(error3);
        Serial.print(" w31':");
        Serial.print(w31);
        Serial.print(" w32':");
        Serial.print(w32);
        Serial.print(" b3':");
        Serial.print(bias3);
        Serial.println(" ");
        }

     if(i == (trng_set_size-1))        //end of Epoch
        {
        MSE3 = sq_error3/trng_set_size;
        Serial.println(" ");
        Serial.print("MSE3:");
        Serial.print(MSE3);
        Serial.println(" ");
        }

      Serial.println(" ");
      }//end for
    }//if (MSE3)

  }
else                                    //All MSEs = 0
  {                                     //Run Mode
  Serial.println("else...Run Mode");

  //print results
  Serial.println(" ");
  Serial.print(" w11':");
  Serial.print(w11);
  Serial.print(" w12':");
  Serial.print(w12);
  Serial.print(" b1':");
  Serial.print(bias1);
  Serial.println(" ");

  //print results
  Serial.println(" ");
  Serial.print(" w21':");
```

```
Serial.print(w21);
Serial.print(" w22':");
Serial.print(w22);
Serial.print(" b2':");
Serial.print(bias2);
Serial.println(" ");

//print results
Serial.println(" ");
Serial.print(" w31':");
Serial.print(w31);
Serial.print(" w32':");
Serial.print(w32);
Serial.print(" b3':");
Serial.print(bias3);
Serial.println(" ");

//flush input buffer
while(Serial.available() >0)
  {
  Serial.read();
  }
//request x1 input value from user via serial monitor
Serial.println("Insert new value of x1: [send]");
while(Serial.available()==0){}        //wait for user input data
x1 = Serial.parseFloat();
Serial.println(" ");
Serial.print("x1:");
Serial.println(x1);

//flush input buffer
while(Serial.available() >0)
  {
  Serial.read();
  }
//request x input value from user via serial monitor
Serial.print("Insert new value of x2: [send]");
delay(5000);
while(Serial.available()==0){}        //wait for user input data
x2 = Serial.parseFloat();
Serial.println(" ");
Serial.print("x2:");
Serial.println(x2);

//process new input and assign to appropriate category
//illuminate appropriate LED
net1 = (x1*w11) + (x2*w12) + bias1;
net2 = (x1*w21) + (x2*w22) + bias2;
net3 = (x1*w31) + (x2*w32) + bias3;
if(net1 > 0)
  {
  //category 1
  Serial.println("Category 1");
```

```
    Serial.print("net1: ");
    Serial.println(net1);
    }
  else if (net2 > 0)
    {
    //category 2
    Serial.println("Category 2");
    Serial.print("net2: ");
    Serial.println(net2);
    }
  else if (net3 > 0)
    {
    //category 3
    Serial.println("Category 3");
    Serial.print("net3: ");
    Serial.println(net3);
    }
  else
    {
    //Error
    Serial.println("Category Error");
    }
  delay(5000);
  }
}


//*****************************************************************
```

6.4.1   Three Perceptron Run Mode

This section provides the "Run Mode" for the three perceptron model explored in the previous section. We have streamlined the sketch by removing storage to EEPROM, use of an external "Train/Run Mode" DIP switch, and the LED status indicators. Also, we have tested the sketch on both the Arduino UNO R3 and also the Arduino Nano 33 BLE sense. The Run Mode begins at the "else" statement in the sketch above.

To test the sketch, compile and upload the sketch to either an Arduino Uno R3 or the Arduino Nano BLE 33 Sense. When the training portion of the sketch is complete, it enters the run mode. You will be prompted for values of x1 and x2. Use values both inside and outside the original training set. Verify the algorithm places the data into the proper category.

6.5   Perceptron Challenges

Perceptrons are useful for separating input/output data pairs that are linearly separable. For data sets not meeting these requirements, advanced techniques are required [5]. For the remainder of the chapter we explore artificial neural networks (ANNs).

6.6   Artificial Neural Network (ANN)

In this section we explore the Artificial Neural Network or ANN. We begin with a discussion of the fundamental building block neuron model. We then use the neuron building block in assembling an ANN containing several layers of interacting neurons. We investigate a feed forward ANN and then discuss the concept of backpropagation. The ANN is then modified to include backpropagation features. Along the way we provide Arduino sketches to implement the neuron, a feed forward ANN, and an ANN with backpropagation features. The mathematical description here is based on the excellent development provided in [4, 6, 8].

6.6.1   Single Neuron Model

The model of a single neuron is shown in Fig. 6.12. Note the striking similarity between the neuron and the perceptron models. We modify the perceptron model by changing the activation function to the sigmoid function and also the equations for updating the weights and bias as shown.

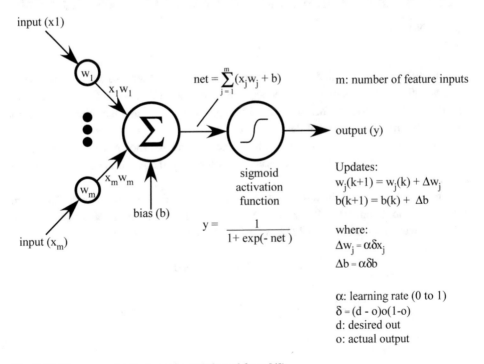

input (x1)

$$net = \sum_{j=1}^{m}(x_j w_j + b)$$

m: number of feature inputs

w_1

$x_1 w_1$

output (y)

$x_m w_m$

w_m

bias (b)

sigmoid
activation
function

$$y = \frac{1}{1 + exp(-net)}$$

input (x_m)

Updates:
$w_j(k+1) = w_j(k) + \Delta w_j$
$b(k+1) = b(k) + \Delta b$

where:
$\Delta w_j = \alpha \delta x_j$
$\Delta b = \alpha \delta b$

α: learning rate (0 to 1)
$\delta = (d - o)o(1-o)$
d: desired out
o: actual output

Fig. 6.12 Neuron model for input layer (adapted from [4])

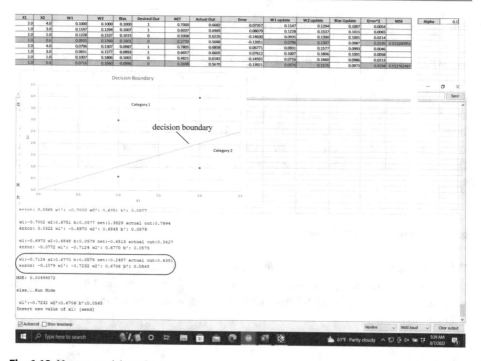

Fig. 6.13 Neuron model results

It is important to note that the model converges and the mean squared error (MSE) declines with the processing of each epoch. However, the MSE may not reach a value of zero after multiple epochs. Therefore, the sketch has features to set a maximum number of iterations or a minimum value of desired MSE.

As before, we verify the equations with an Excel model and then proceed to testing the Arduino sketch. The Excel model results and the test run for the Arduino sketch is provided in Fig. 6.13.

To provide the decision boundary between category 1 and 2 data, the sketch was run on an Arduino Uno R3 until the MSE was less than 0.005. This required 41 epochs and resulted in the following values: w1 = −0.7232, w2 = 0.6706, and bias = 0.0568.

$$x1w1 + x2w2 + b = 0$$

$$x1(-0.7232) + x2(0.6706) + (0.0568) = 0$$

$$y = mx + b$$

$$X2 = 1.08x1 - 0.08$$

This separating line is shown in Fig. 6.13.

```
//*****************************************************************
//neuron_2in_1out_train_run
//Sketch has two modes of operation: Train and Run
//Model first performs training and then proceeds to run.
//*****************************************************************
//define
#define trng_set_size        4 //entries in training set

//global
double input_1[trng_set_size]        = {2.0, 1.0, 2.0, 1.0};
double input_2[trng_set_size]        = {4.0, 3.0, 1.0, 0.6};
double desired_output[trng_set_size] = {1.0, 1.0,   0,   0};
double actual_output;
unsigned int epoch_iterations = 10000;
unsigned int epoch_count = 0;
double w1   = 0.10;              //set initial weight of w1
double w2   = 0.10;              //set initial weight of w2
double bias = 0.10;             //set inital bias value
double learning_rate = 0.1;     //set 0 to 1
double x1, x2;                  //neuron inputs
double error;                   //desired vs actual output
double net;                     //numerical out from neuron
double sq_error = 0;            //squared error
double MSE = 1;                 //mean squared error
double MSE_old=1, MSE_new=2;    //detect change in MSE
double MSE_goal = 0.005;        //MSE goal


void setup()
{
Serial.begin(9600);            //Set the Serial output
epoch_count = 0;               //reset epoch count
}

void loop()
{
int i;
while(!Serial);
                               //train until condition not met
if((MSE > 0)&&(epoch_count <=epoch_iterations)&&(MSE_old != MSE_new)&&(MSE_new > MSE_goal))
  {
  epoch_count = epoch_count + 1;//epoch count
  Serial.print("Epoch:");
  Serial.println(epoch_count);

  net = 0;                     //reset net
  sq_error = 0;                //reset sq error
  for(i=0; i < trng_set_size; i++)
    {
    net = (input_1[i]*w1) + (input_2[i]*w2) + bias;
    actual_output = (double) (1.0/(1.0 + exp(-net)));

    Serial.print(" w1:");
    Serial.print(w1, 4);
    Serial.print(" w2:");
    Serial.print(w2, 4);
    Serial.print(" b:");
    Serial.print(bias, 4);
```

```
Serial.print(" net:");
Serial.print(net, 4);
Serial.print(" actual out:");
Serial.println(actual_output, 4);

error = (desired_output[i] - actual_output) * (actual_output) * (1 - actual_output);
sq_error = sq_error + (error * error);
w1 = w1 + (learning_rate * error * input_1[i]);
w2 = w2 + (learning_rate * error * input_2[i]);
bias = bias + (learning_rate * error * bias);

Serial.print(" error: ");
Serial.print(error, 4);
Serial.print(" w1': ");
Serial.print(w1, 4);
Serial.print(" w2': ");
Serial.print(w2, 4);
Serial.print(" b': ");
Serial.print(bias, 4);
Serial.println(" ");

if(i == (trng_set_size-1))      //end of Epoch
  {
  MSE_old = MSE;
  MSE = sq_error/trng_set_size;
  Serial.println(" ");
  Serial.print("MSE: ");
  Serial.print(MSE, 8);
  Serial.println(" ");
  MSE_new = MSE;
  }

Serial.println(" ");
}//end for
}//if (MSE)

else                                    //run mode
 {
 Serial.println("else...Run Mode");

 //print results
 Serial.println(" ");
 Serial.print(" w1':");
 Serial.print(w1, 4);
 Serial.print(" w2':");
 Serial.print(w2, 4);
 Serial.print(" b':");
 Serial.print(bias, 4);
 Serial.println(" ");

 //flush input buffer
 while(Serial.available() >0)
   {
   Serial.read();
   }
 //request x1 input value from user via serial monitor
 Serial.println("Insert new value of x1: [send]");
 while(Serial.available()==0){}          //wait for user input data
 x1 = Serial.parseFloat();
 Serial.println(" ");
 Serial.print("x1:");
 Serial.println(x1, 4);

 //flush input buffer
```

```
while(Serial.available() >0)
  {
  Serial.read();
  }
//request x input value from user via serial monitor
Serial.print("Insert new value of x2: [send]");
delay(5000);
while(Serial.available()==0){}         //wait for user input data
x2 = Serial.parseFloat();
Serial.println(" ");
Serial.print("x2:");
Serial.println(x2, 4);

//process new input and assign to appropriate category
//illuminate appropriate LED
net = (x1*w1) + (x2*w2) + bias;
actual_output = (double) (1.0/(1.0 + exp(-net)));
Serial.print("Actual out: ");
Serial.println(actual_output);
if(actual_output >= 0.5)
  Serial.println("Category 1");
else
  Serial.println("Category 2");
delay(5000);
}//end else
}

//**************************************************************
```

It is important to carefully select the training set for the ANN. Once training is complete the neuron may be used to process input/output pairs outside the training set. We call this the "Run Mode." We investigate this mode in the next section.

6.6.2   Single Neuron Run Mode

The "Run Mode" for the single neuron model explored in the last section begins at the "else" statement. To test the sketch, compile and upload the sketch to either an Arduino UNO R3 or the Arduino Nano BLE 33 Sense. When the training portion of the sketch is complete, it enters the Run Mode. The user is prompted for values of feature inputs x1 and x2. Use values both inside and outside the original training set. Verify the algorithm places the data into the proper category.

6.6.3   Artificial Neural Networks

The individual neuron serves as the main building block for an artificial neural network (ANN) as shown in Fig. 6.14a. This specific ANN consists of three layers: the L1 input layer, the L2 hidden layer, and the L3 output layer. Each neuron is shown as a node (circle) in the figure.

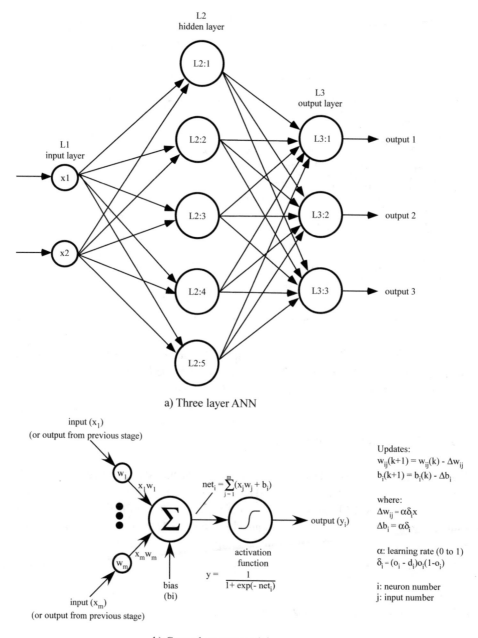

a) Three layer ANN

b) General neuron model.

Fig. 6.14 a Three layer ANN model, **b** generalized single neuron model (adapted from [4])

The weighted inputs (or features) to the L1 layer are routed to each node (neuron) in the L2 hidden layer. It is called the hidden layer since it has no direct external inputs or outputs to the ANN. The output from nodes in Layer 2 serve as inputs to nodes in Layer 3. The illustrated ANN shows a fully connected network of nodes. That is, each node output in a given later provides input to every node in the next layer. Nodes in Layers 2 and 3 are modeled using the general model shown in Fig. 6.14b.

To help track the nomenclature used to label node (neuron) inputs and outputs an alternative scheme is used as shown in Fig. 6.16. In this simplified line diagram the nodes (neurons) are shown as vertical lines. The corresponding inputs, bias, and output for each node are shown.

The ANN is analyzed in the forward direction from left to right using the neuron model provided in Fig. 6.14b and the procedural notes provided in Fig. 6.15. The step–by–step approach for analyzing the ANN is illustrated in Fig. 6.16. The numbers within the "bubbles" correspond to the corresponding steps in Fig. 6.15. The mathematical description provided here is based on the excellent development by [4, 6, 8].

The error terms developed at the ANN outputs are backpropagated from right to left to update the weights of the other layers as shown in Fig. 6.17. The overall goal is to adjust the ANN network weights and biases to reduce the error between actual and desired outputs. As shown in red, the errors at each output are backpropagated through the ANN network to update the upstream weights.

As before, a training set is used to iteratively set the weights and biases until desired error values at the ANN outputs are achieved. The ANN network weights and biases may be updated after each input/output pair sample, after a batch of some set number of input/output samples, or after the entire set of samples (epoch) is completed. In the upcoming sketch, we update the ANN network weights after each input/output sample [2].

It is important to carefully select the training set for the ANN. Once training is complete the ANN may be used to process input/output pairs outside the training set. We call this the "Run Mode." If a new input/output pair entered into the ANN that was well characterized by the training set, the ANN will operate in the interpolation mode. On the other hand, if a new input/output pair is chosen in a poorly characterized area, the ANN operates in the extrapolation mode and may yield poor results [10]. For this specific example, the training set shown in Fig. 6.18 is used. A high level UML activity diagram is provided in Fig. 6.19 for the ANN sketch.

```
//**********************************************************************
//ANN_2in_3layer_3out_with_bias_max_mse_random_z_score
//Sketch has two modes of operation: Train and Run
//Model first performs training and then proceeds to run.
//ANN model has the following characteristics:
//   - 2 inputs
//   - 3 layers
//   - 3 outputs
//   - each node (neuron) has bias input
```

Notes:
1. Node designation: Llayer#:neuron#
2. Number of neurons in Layers L1, L2, and L3 - n, m, l
3. Weight designation: w_{lij} $w_{layer\#neuron\#input\#}$

Training process:
1. Initialize all weights and biases in all layers to small random values.
2. Provide training set input to x1 and x2 at Layer L1.
3. Output calculation proceeds layer-to-layer.
  - Start at layer L2, calculate L2 net value for each neuron.
  - Process L2 net values to activation functions to determine L2 layer neuron outputs.
  - Layer 2 outputs now serve as Layer 3 inputs.
  - Calculate L3 net value for each neuron.
  - Process L3 net values to activation functions to determine L3 layer neuron outputs.

$$net_i = \sum_{j=1}^{m} x_{ij}w_j + b \qquad O_i = \frac{1}{1+ \exp(-net_i)}$$

4. Calculate change in weights and biases between Layers 2 and 3
  - Compare Layer 3 actual output (O) to desired output (d).
  - Calculate change in weight: $\Delta w_{ij} = \alpha \delta o_j$, $\Delta b_i = \alpha \delta_i$

  α: learning rate (0 to 1) $\delta_i = (o_i - d_i)o_i(1-o_i)$

  o_j: output of neuron j in Layer L2

5. Calculate change in weights and biases between Layers 1 and 2: $\Delta w_{ij} = \beta \delta_{HI}o_j$, $\Delta b_i = \beta \delta_{HI}$

  $\delta_{HI} = o_i(1 - o_i)\sum_{k=1}^{m} \delta_k w_{ik}$ β: learning rate (0 to 1)

  o_j: output of neuron j in Layer L1

  o_i: output of neuron i in Layer L2

  δ_k: from neurons in Layer L3

  w_{ik}: weight connecting a specific neuron (k) in the L2 layer
          to a specific neuron (i) in the L3 layer

6. Update weights and biases between Layers L2 and L3 and Layers L1 and L2:

  $w_{ij}(k+1) = w_{ij}(k) - \Delta w_{ij}$, $b_i(k+1) = b_i(k) - \Delta b_i$

7. Obtain the error for neurons in Layer L3:

  $$E = \frac{1}{2}\sum_{i=1}^{L} (o_i - d_i)^2$$

8. If error term has reached desired goal, stop training
  else continue training.

Fig. 6.15 ANN analysis process (Kulkarni)

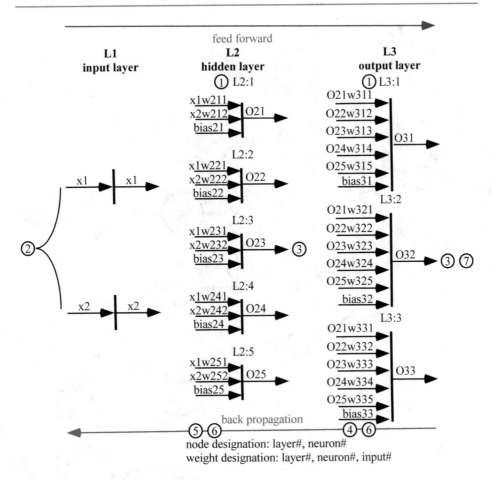

Fig. 6.16 Simplified ANN representation

```
//  - fully connected layers
//*********************************************************************

//define
#define trng_set_size      24    //entries in training set
#define troubleshoot        0    //flag to turn prints on(1)/off(0)

//data sets input_x1, input_x2/desired_output31, 32, 33
float input_x1[trng_set_size]  = {25.0, 12.0, 20.0,  1.0, 20.0, 12.0,
                                  15.0,  5.0, 80.0, 87.0, 95.0, 76.0,
                                  85.0, 87.0, 90.0, 70.0, 40.0, 57.0,
                                  35.0, 26.0, 30.0, 55.0, 25.0, 20.0};
float input_x2[trng_set_size]  = {62.0, 75.0, 80.0, 79.0, 62.0, 55.0,
                                  85.0, 85.0, 87.0, 80.0, 95.0, 94.0,
                                  85.0, 80.0, 90.0, 90.0, 17.0, 20.0,
```

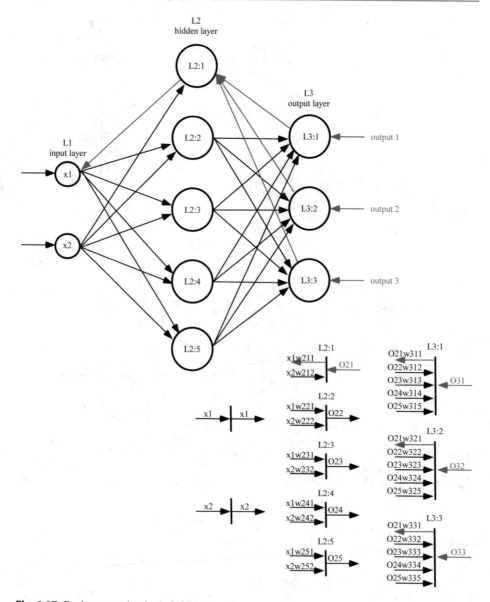

Fig. 6.17 Backpropagation in the ANN network

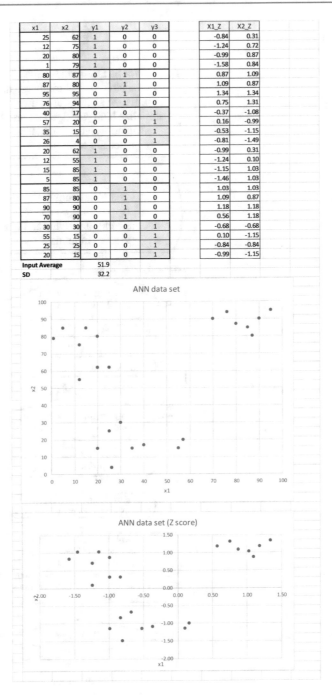

x1	x2	y1	y2	y3		X1_Z	X2_Z
25	62	1	0	0		-0.84	0.31
12	75	1	0	0		-1.24	0.72
20	80	1	0	0		-0.99	0.87
1	79	1	0	0		-1.58	0.84
80	87	0	1	0		0.87	1.09
87	80	0	1	0		1.09	0.87
95	95	0	1	0		1.34	1.34
76	94	0	1	0		0.75	1.31
40	17	0	0	1		-0.37	-1.08
57	20	0	0	1		0.16	-0.99
35	15	0	0	1		-0.53	-1.15
26	4	0	0	1		-0.81	-1.49
20	62	1	0	0		-0.99	0.31
12	55	1	0	0		-1.24	0.10
15	85	1	0	0		-1.15	1.03
5	85	1	0	0		-1.46	1.03
85	85	0	1	0		1.03	1.03
87	80	0	1	0		1.09	0.87
90	90	0	1	0		1.18	1.18
70	90	0	1	0		0.56	1.18
30	30	0	0	1		-0.68	-0.68
55	15	0	0	1		0.10	-1.15
25	25	0	0	1		-0.84	-0.84
20	15	0	0	1		-0.99	-1.15
Input Average		51.9					
SD		32.2					

Fig. 6.18 Normalized ANN input data

Fig. 6.19 High level ANN
UML activity diagram

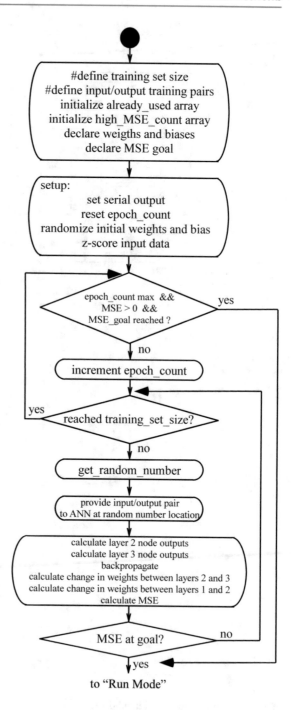

to "Run Mode"

```
                                   15.0,   4.0, 30.0, 15.0, 25.0, 15.0};
int desired_output31[trng_set_size] = {1, 1, 1, 1, 1, 1, 1, 1,
                                        0, 0, 0, 0, 0, 0, 0, 0,
                                        0, 0, 0, 0, 0, 0, 0, 0};
int desired_output32[trng_set_size] = {0, 0, 0, 0, 0, 0, 0, 0,
                                        1, 1, 1, 1, 1, 1, 1, 1,
                                        0, 0, 0, 0, 0, 0, 0, 0};
int desired_output33[trng_set_size] = {0, 0, 0, 0, 0, 0, 0, 0,
                                        0, 0, 0, 0, 0, 0, 0, 0,
                                        1, 1, 1, 1, 1, 1, 1, 1};
unsigned int already_used[trng_set_size] =   {0,0,0,0,0,0,0,0,0,0,0,0,
                                        0,0,0,0,0,0,0,0,0,0,0,0};
unsigned int high_MSE_count[trng_set_size] = {0,0,0,0,0,0,0,0,0,0,0,0,
                                        0,0,0,0,0,0,0,0,0,0,0,0};
unsigned int epoch_iterations = 10000;    //maximum number of iterations
unsigned int epoch_count = 0;             //tracks number of interations

//layer 3 actual outputs
double actual_output31, actual_output32, actual_output33;

//layer 2 weights and biases
double w211, w212, w221, w222, w231;
double w232, w241, w242, w251, w252;
double bias21, bias22, bias23, bias24, bias25;

//layer 3 weights and biases
double w311, w312, w313, w314, w315;
double w321, w322, w323, w324, w325;
double w331, w332, w333, w334, w335;
double bias31,  bias32, bias33;

double learning_rate = 0.05;                     //set from0 to 1
double x1, x2;                                    //neuron inputs
double delta21, delta22, delta23, delta24, delta25;
double delta31, delta32, delta33;    //desired vs actual output
double net21, net22,  net23, net24, net25; //Layer 2 net output
double net31, net32,  net33;             //Layer 3 net output
double out21, out22,  out23, out24, out25;    //Layer 2 output
double out31, out32, out33;              //Layer 3 output
double sq_error31= 0, sq_error32= 0, sq_error33 = 0; //sq error
double MSE = 1;                          //data point MSE
double MSE_goal = 0.005;                          //MSE goal
double max_MSE_epoch = 1;      //tracks max MSE per data set run
float input_average, input_stddev; //used for input z transform

void setup()
{
Serial.begin(9600);                    //set the Serial output
epoch_count = 0;                       //reset epoch count
randomize_weights_bias();         //randomize weights and bias
z_score_input();                   //z score inputs x1 and x2
}
```

```
void loop()
{
int i, j, m;
unsigned new_number_selected;

while(!Serial);                               //wait for serial connection
                                   //train until condition not met
if((epoch_count<=epoch_iterations)&&(MSE>0)&&(max_MSE_epoch>MSE_goal))
  {
  max_MSE_epoch = -1;                                   //reset value
  epoch_count = epoch_count + 1;          //increment epoch count
  if(troubleshoot)
    {
    Serial.print("Epoch: ");
    Serial.print(epoch_count);
    }
  initialize_array();              //initialize random number used
                                   //start training epoch
  for(j=0; j < trng_set_size; j++)
    {
    i = get_unused_random_number();
    if(troubleshoot)
      {
      Serial.print(" ");
      Serial.print(i);
      }
    //with random number "i" selected, analyze ANN with i data set
    //Calculate Layer 2 neuron outputs
    net21 = (input_x1[i]*w211)+(input_x2[i]*w212)+bias21;//L2,N1
    out21 = (double) (1.0/(1.0 + exp(-net21)));
    net22 = (input_x1[i]*w221)+(input_x2[i]*w222)+bias22;//L2,N2
    out22 = (double) (1.0/(1.0 + exp(-net22)));
    net23 = (input_x1[i]*w231)+(input_x2[i]*w232)+bias23;//L2,N3
    out23 = (double) (1.0/(1.0 + exp(-net23)));
    net24 = (input_x1[i]*w241)+(input_x2[i]*w242)+bias24;//L2,N4
    out24 = (double) (1.0/(1.0 + exp(-net24)));
    net25 = (input_x1[i]*w251)+(input_x2[i]*w252)+bias25;//L2,N5
    out25 = (double) (1.0/(1.0 + exp(-net25)));

    //Layer 2 outputs become inputs to Layer 3
    //Calculate Layer 3 neuron outputs
    net31 = (out21*w311)+(out22*w312)+(out23*w313)+        //L3,N1
            (out24*w314)+(out25*w315) + bias31;
    out31 = (double) (1.0/(1.0 + exp(-net31)));
    net32 = (out21*w321)+(out22*w322)+(out23*w323)+        //L3,N2
            (out24*w324)+(out25*w325) + bias32;
    out32 = (double) (1.0/(1.0 + exp(-net32)));
    net33 = (out21*w331)+(out22*w332)+(out23*w333)+        //L3,N3
            (out24*w334)+(out25*w335) + bias33;
    out33 = (double) (1.0/(1.0 + exp(-net33)));

    //Calculate change in weights between Layers 2 and 3
    //neuron Layer 3, Neuron 1 (L3:1)
```

```
delta31=(out31 - desired_output31[i])*(out31)*(1-out31);
w311 = w311 - (learning_rate * delta31 * out21);
w312 = w312 - (learning_rate * delta31 * out22);
w313 = w313 - (learning_rate * delta31 * out23);
w314 = w314 - (learning_rate * delta31 * out24);
w315 = w315 - (learning_rate * delta31 * out25);
bias31 = bias31 - (learning_rate * delta31);

//neuron Layer 3, Neuron 2 (L3:2)
delta32=(out32 - desired_output32[i])*(out32)*(1-out32);
w321 = w321 - (learning_rate * delta32 * out21);
w322 = w322 - (learning_rate * delta32 * out22);
w323 = w323 - (learning_rate * delta32 * out23);
w324 = w324 - (learning_rate * delta32 * out24);
w325 = w325 - (learning_rate * delta32 * out25);
bias32 = bias32 - (learning_rate * delta32);

//neuron Layer 3, Neuron 3 (L3:3)
delta33=(out33 - desired_output33[i])*(out33)*(1-out33);
w331 = w331 - (learning_rate * delta33 * out21);
w332 = w332 - (learning_rate * delta33 * out22);
w333 = w333 - (learning_rate * delta33 * out23);
w334 = w334 - (learning_rate * delta33 * out24);
w335 = w335 - (learning_rate * delta33 * out25);
bias33 = bias33 - (learning_rate * delta33);

//Calculate change in weights between Layers 1 and 2
//neuron Layer 2, Neuron 1 (L2:1)
delta21 = out21 * (1 - out21) *
          ((w311*delta31)+(w321*delta32)+(w331*delta33));
w211 = w211 - (learning_rate * input_x1[i] * delta21);
w212 = w212 - (learning_rate * input_x2[i] * delta21);
bias21 = bias21 - (learning_rate * delta21);

//neuron Layer 2, Neuron 2 (L2:2)
delta22 = out22 * (1 - out22) *
          ((w312*delta31)+(w322*delta32)+(w332*delta33));
w221 = w221 - (learning_rate * input_x1[i] * delta22);
w222 = w222 - (learning_rate * input_x2[i] * delta22);
bias22 = bias22 - (learning_rate * delta22);

//neuron Layer 2, Neuron 3 (L2:3)
delta23 = out23 * (1 - out23) *
          ((w313*delta31)+(w323*delta32)+(w333*delta33));
w231 = w231 - (learning_rate * input_x1[i] * delta23);
w232 = w232 - (learning_rate * input_x2[i] * delta23);
bias23 = bias23 - (learning_rate * delta23);

//neuron Layer 2, Neuron 4 (L2:4)
delta24 = out24 * (1 - out24) *
          ((w314*delta31)+(w324*delta32)+(w334*delta33));
w241 = w241 - (learning_rate * input_x1[i] * delta24);
w242 = w242 - (learning_rate * input_x2[i] * delta24);
```

```
   bias24 = bias24 - (learning_rate * delta24);

   //neuron Layer 2, Neuron 5 (L2:5)
   delta25 = out25 * (1 - out25) *
            ((w315*delta31)+(w325*delta32)+(w335*delta33));
   w251 = w251 - (learning_rate * input_x1[i] * delta25);
   w252 = w252 - (learning_rate * input_x2[i] * delta25);
   bias25 = bias25 - (learning_rate * delta25);

   //Calculate error at the end of each sample
   MSE=0.5*(((desired_output31[i]-out31)*(desired_output31[i]-out31))+
            ((desired_output32[i]-out32)*(desired_output32[i]-out32))+
            ((desired_output33[i]-out33)*(desired_output33[i]-out33)));
   if(troubleshoot)
      {
      Serial.print("MSE: ");
      Serial.println(MSE, 4);
      }

   //Print max MSE at the end of an epoch
   if(MSE > max_MSE_epoch)
      {
      max_MSE_epoch = MSE;
      m = i;
      if(troubleshoot)
         {
         Serial.print("max_MSE_epoch updated");
         Serial.println(max_MSE_epoch, 4);
         }
      }

  if(j == (trng_set_size-1))      //end of Epoch
      {
      Serial.print("Epoch: ");
      Serial.print(epoch_count);
      Serial.print("    max i : ");
      Serial.print(m);
      Serial.print("    max_MSE_epoch: ");
      Serial.println(max_MSE_epoch, 4);

      high_MSE_count[m] = high_MSE_count[m] + 1;
      }
   }//end for
 }//if (MSE)

else                                      //run mode
  {
  if(troubleshoot)
    {
    Serial.print("Epoch: ");
    Serial.print(epoch_count);

    for(j=0; j < trng_set_size; j++)
```

```
    {
    Serial.print("Training set: ");
    Serial.print(j);
    Serial.print("   Training set count: ");
    Serial.println(high_MSE_count[j]);
    }
  }

//flush input buffer
while(Serial.available() >0)
  {
  Serial.read();
  }
//request x1 input value from user via serial monitor
Serial.println("Insert new value of x1: [send]");
while(Serial.available()==0){}              //wait for user input data
x1 = Serial.parseFloat();
x1 = (x1 - input_average)/input_stddev; //z: input x1
Serial.println(" ");
Serial.print("x1:");
Serial.println(x1, 4);

//flush input buffer
while(Serial.available() >0)
  {
  Serial.read();
  }
//request x input value from user via serial monitor
Serial.print("Insert new value of x2: [send]");
delay(5000);
while(Serial.available()==0){}              //wait for user input data
x2 = Serial.parseFloat();
x2 = (x2 - input_average)/input_stddev; //z: input x2
Serial.println(" ");
Serial.print("x2:");
Serial.println(x2, 4);

//process new input and assign to appropriate category
//Calculate Layer 2 neuron outputs
net21 = (x1*w211) + (x2*w212) + bias21;     //layer 2, node1
out21 = (double) (1.0/(1.0 + exp(-net21)));

net22 = (x1*w221) + (x2*w222) + bias22;     //layer 2, node2
out22 = (double) (1.0/(1.0 + exp(-net22)));

net23 = (x1*w231) + (x2*w232) + bias23;     //layer 2, node3
out23 = (double) (1.0/(1.0 + exp(-net23)));

net24 = (x1*w241) + (x2*w242) + bias24;     //layer 2, node4
out24 = (double) (1.0/(1.0 + exp(-net24)));

net25 = (x1*w251) + (x2*w252)+ bias25;      //layer 2, node5
out25 = (double) (1.0/(1.0 + exp(-net25)));
```

```
//Layer 2 outputs become inputs to Layer 3
//Calculate Layer 3 neuron outputs
net31 = (out21*w311)+(out22*w312)+(out23*w313)+
        (out24*w314)+(out25*w315) + bias31;
out31 = (double) (1.0/(1.0 + exp(-net31)));

net32 = (out21*w321)+(out22*w322)+(out23*w323)+
        (out24*w324)+(out25*w325) + bias32;
out32 = (double) (1.0/(1.0 + exp(-net32)));

net33 = (out21*w331)+(out22*w332)+(out23*w333)+
        (out24*w334)+(out25*w335) + bias33;
out33 = (double) (1.0/(1.0 + exp(-net33)));

Serial.print("out31: ");
Serial.print(out31);
Serial.print(" out32: ");
Serial.print(out32);
Serial.print(" out33: ");
Serial.println(out33);

delay(5000);
}//end else
}


//*********************************************************************

void randomize_weights_bias()
{
//layer 2 weights and biases
w211   = get_random_weight(); w212   = get_random_weight();
w221   = get_random_weight(); w222   = get_random_weight();
w231   = get_random_weight(); w232   = get_random_weight();
w241   = get_random_weight(); w242   = get_random_weight();
w251   = get_random_weight(); w252   = get_random_weight();
bias21 = get_random_weight(); bias22 = get_random_weight();
bias23 = get_random_weight(); bias24 = get_random_weight();
bias25 = get_random_weight();

//layer 3 weights and biases
w311   = get_random_weight(); w312   = get_random_weight();
w313   = get_random_weight(); w314   = get_random_weight();
w315   = get_random_weight(); w321   = get_random_weight();
w322   = get_random_weight(); w323   = get_random_weight();
w324   = get_random_weight(); w325   = get_random_weight();
w331   = get_random_weight(); w332   = get_random_weight();
w333   = get_random_weight(); w334   = get_random_weight();
w335   = get_random_weight(); bias31 = get_random_weight();
bias32 = get_random_weight(); bias33 = get_random_weight();
}


//*********************************************************************
```

```
double get_random_weight()
{
double          random_input;
unsigned int random_min = 1, random_max = 21;
unsigned int scale_factor = 200.0;

randomSeed(analogRead(A0));
random_input = ((double)(random(random_min, random_max)))/scale_factor;
if(random(0,101) > 50)                      //randommize sign
   random_input = random_input * -1.0;
return random_input;
}

//****************************************************************

void initialize_array(void)
{
unsigned int k;

for(k=0; k < trng_set_size; k++)                       //initialize array
  already_used[k] = 0;
}

//****************************************************************

unsigned int get_unused_random_number(void)
{
unsigned rn;
unsigned new_number_selected;

new_number_selected = 0;
while(new_number_selected != 1)
  {
  rn = (unsigned int) (random(0, trng_set_size));
  if(already_used[rn] == 0)   //number not used yet
    {
    already_used[rn] = 1;     //number has now been used
    new_number_selected = 1;
    }
  else
    {
    new_number_selected = 0;
    }
  }//end while

return rn;
}

//****************************************************************

void z_score_input()
{
```

```
int i;

input_average = 0;

for(i=0; i< trng_set_size; i++)
  {
  input_average = input_average + input_x1[i];
  }
for(i=0; i< trng_set_size; i++)
  {
  input_average = input_average + input_x2[i];
  }

input_average = input_average/(2.0 * trng_set_size);

Serial.print("Average: ");
Serial.println(input_average);

input_stddev = 0;

for(i=0; i< trng_set_size; i++)
  {
  input_stddev = input_stddev + ((input_x1[i] - input_average) *
                                 (input_x1[i] - input_average));
  }
for(i=0; i< trng_set_size; i++)
  {
  input_stddev = input_stddev + ((input_x2[i] - input_average) *
                                 (input_x2[i] - input_average));
  }

input_stddev = input_stddev/((2.0 * trng_set_size) - 1);

input_stddev = sqrt(input_stddev);

/*Serial.print("Input stddev: ");
Serial.println(input_stddev); */

//Generate x1 Z scores
for(i=0; i< trng_set_size; i++)
  {
  input_x1[i] = (input_x1[i] - input_average)/input_stddev;
  if(troubleshoot)
    {
    Serial.print("x1: ");
    Serial.print(i);
    Serial.print("    Input x1 z score: ");
    Serial.println(input_x1[i]);
    }
  }

//Generate x2 Z scores
for(i=0; i< trng_set_size; i++)
```

```
{
input_x2[i] = (input_x2[i] - input_average)/input_stddev;
if(troubleshoot)
  {
  Serial.print("x2: ");
  Serial.print(i);
  Serial.print("     Input x2 z score: ");
  Serial.println(input_x2[i]);
  }
}
}

//****************************************************************
```

6.6.4   ANN Convergence

During ANN training, the mean squared error should decline and eventually reach the desired goal. Depending on the specific ANN configuration and data set, this may not occur. If an ANN does not converge toward the MSE goal, the following actions may be taken:

- Randomize initial weights and biases to small numbers. In the previous sketch, this is accomplished by function "randomize_weights_bias()." Within the function each ANN weight and bias is initially set to a small random value ranging from ±0.005 to 0.100
- Randomize the order that input/output data pairs are provided to the ANN in a given epoch. In the previous sketch, this is accomplished by function "get_unused_random_number()." This function provides a random number from 0 to the training_set_size–1 to select a random order of data set input/output pairs to the ANN for training.
- Provide a larger ANN training set. It is recommended to have ten different entries for each input feature. In the previous sketch we used 24 input/output data pairs for the two input features x1 and x2.
- Precondition the data set. In the previous sketch, this is accomplished by function "z_score_input()." This data normalization set replaces each input in the data set with its z score equivalent [1]. A given input value's z score is calculated by subtracting the data set input mean from the input value and then dividing by the standard deviation of the input data set. This has the overall effect of providing an equivalent data set with a mean of zero and a standard deviation of one as shown in Fig. 6.18.
- Adjust the ANN architecture (e.g. change the number of nodes within the hidden layer). Several authors have indicated this is a trial and error process.

6.7   Deep Neural Networks–Introduction to Software Tools

In this chapter we have explored the fundamentals of perceptrons and ANNs. These basic concepts may be extended to develop multilayer or deep neural networks. To explore these concepts require advanced machine learning tools based on the TensorFlow software suite. The tool suite was originally developed by Google and now is open source software.

In this section we provide a brief introduction to the TensorFlow software tool suite. This is a logical next step in your study of AI and ML concepts. The reader is highly encouraged to obtain a copy of "TinyML–Machine Learning with TensorFlow Lite on Arduino and Ultra–Low–Power Microcontrollers" by Pete Warden and Daniel Situnayake to pursue this next step.[1] The book provides a thorough, step–by–step introduction to advanced machine learning tools and deep learning network development.

The deep learning workflow used by TensorFlow is shown in Fig. 6.20 [9]. It is similar to the flow used to develop perceptron and ANN models earlier in the chapter. Software tools used in deep learning network include:

- Python: programming language used for ML development,
- TensorFlow: tool suite for training, implementing, and testing a deep learning network,
- TensorFlow Lite: TensorFlow variant for mobile app development,
- TensorFlow Lite for Microcontrollers: TensorFlow variant for microcontroller–based application development[2]; and
- Jupyter Notebooks: provides for a well–documented ML application with supporting comments, code, and visualizations;
- Colaboratory (CoLab): online environment to run and share Jupyter Notebook ML projects.

Typically when developing a deep learning application for a microcontroller steps 1 through 5 are performed by a host computer. Once the model is trained and complete it is loaded to the microcontroller for inference testing and refinement. Warden and Situnayake provide four trained models in their "Arduino_TensorFlowLibrary." The library is updated daily and may be downloaded from their book associated download site. Once downloaded, the library may be imported to the Arduino IDE (Sketch– >Include Library– >Add .Zip Library). Trained examples include:

- hello_world: The algorithm takes as input a value of x and predicts an output $y = sin(x)$.
- magic_wand: The trained algorithm uses the onboard accelerometer to detect a wing, ring, or slope wand gestures and provide LED and Serial Monitor output.
- micro_speech: Using a trained speech recognition application, illuminates a green LED when the word "yes" is spoken and turns on a red LED when the word "no" is spoken.

---

[1] "TinyML–Machine Learning with TensorFlow Lite on Arduino and Ultra–Low–Power Microcontrollers," Pete Warden and Daniel Situnayake, O'Reilly [9].

[2] Implementation requires a 32–bit microcontroller such as the Arduino Nano 33 BLE Sense.

Fig. 6.20 TensorFlow deep
learning workflow [9]

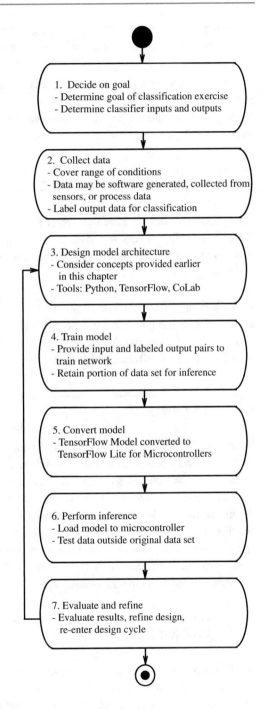

- person_detection: Provides a trained algorithm to take an image as input and determine whether a face is shown or not. If a face is detected an LED is illuminated.

You can see these trained algorithms may be readily adapted to perform other useful and interesting actions. Eventually, you will want to develop and train your own TensorFlow application from scratch. The interested reader is invited to explore this next step in Warden and Situnayake's book.

6.8   Application: ANN Robot Control

In previous chapters we have developed a control algorithm for a maze navigating robot. Develop an ANN network to control a robot with three sensors (left, center, and right) and four possible movement alternatives (left turn, right turn, forward, and reverse) in a maze.
   Deliverables:

- A diagram of your ANN network,
- The data set used to train the ANN network,
- A converging Arduino sketch modeling the ANN network, and
- A test plan demonstrating robot control operation.

6.9   Summary

In this chapter we explored the concept of neurons and neuron models to solve real world challenges. We began with a brief description of the biological neuron and investigated a model of the neuron, called the perceptron, developed by Frank Rosenblatt in 1959. We used the single perceptron model to separate objects into two categories. We extended the model to include additional perceptrons to separate objects into multiple categories. We then investigated the single neuron and then explored the concept of backpropagation and developed a three–layer feed–forward network. Along the way we developed Arduino sketches of these different models. We concluded the chapter with a brief introduction to software tools to develop and implement deep neural networks.

6.10   Problems

1. Describe the capabilities and limitations of a single perceptron model.
2. Describe the capabilities and limitations of a multiple perceptron model.
3. Describe the difference between the train and run modes in an artificial neural network.

4. Implement the Run Mode for the Tomato Sorter example using the single perceptron model.
5. Develop a data set that a multiple perceptron model can not categorize. What limits the model from categorizing the data set?
6. Using the data set developed in the question above demonstrate categorization with an artificial neural network.
7. You have developed an artificial neural network that will not converge. What are possible remedies?
8. How is converge measured in an artificial neural network? Explain.
9. In previous chapters we have developed a control algorithm for a maze navigating robot. Our goal is to develop an ANN network to control a robot with three sensors (left, center, and right) and four possible movement alternatives (left turn, right turn, forward, and reverse) in a maze. Provide a diagram of the ANN network. How many layers will the ANN contain? How many nodes will be in each layer?
10. Develop an Arduino sketch to implement a converging ANN for the control system described in the question above.
11. Compare and contrast fuzzy logic versus ANN mechanisms for controlling a robot.
12. Earlier in the chapter we investigated a tomato sorter application. Implement a tomato sorter to place tomatoes into the following groups: small green, small red, large green, and large red. Will you use a multiple perceptron model or ANN? Explain.
13. Implement the sorter described in the question above. Demonstrate proper sorter operation.
14. What are different techniques for normalizing the input data to an ANN network?
15. Complete the Run Mode for the ANN UML activity diagram.
16. Adapt the micro_speech algorithm in the "Arduino_TensorFlowLibrary" to turn a small motor on and off with voice commands.

References

1. Baeldung, *Normalizing Inputs for an Artificial Neural Network,* www.baeldung.com.
2. J. Brownlee, *Difference Between a Batch and an Epoch in a Neural Network,* www.machinelearningmatery.com, July 2018.
3. Dan, *Single Layer Perceptron Explained,* ML Corner, October 13, 2020.
4. A.D. Kulkarni, *Computer Vision and Fuzzy–Neural Systems,* Prentice Hall, 2001.
5. M.L. Minsky and S.A. Papert, *Perceptrons*, expanded edition, The MIT Press, 1988.
6. T. Rashid, *Make Your Own Neural Network,* Middletown, DE, 2017.
7. C.F. Stevens, *The Neuron,* Scientific American, pp. 55–65, 1979.
8. M. Taylor, *Neural Networks A Visual Introduction for Beginners,* Blue Windmill Media, 2017.
9. P. Warden and D. Situnayake, *TinyML–Machine Learning with TensorFlow Lite on Arduino and Ultra–Low–Power Microcontrollers,* Pete Warden and Daniel Situnayake, O'Reilly, 2020.
10. B.J. Wythoff, *Backpropagation Neural Networks–A Tutorial,* Chemometrics and Intelligent Laboratory Systems, 18 (1993), 115–155.

Index

A
ADC process, 32
American Standard Code for Information
    Interchange (ASCII), 24
Analog-to-Digital Converters (ADC), 33
ANN convergence, 197
ANN extrapolation mode, 183
ANN interpolation mode, 183
ANN model, 181
ANN, z score input, 197
APDS-9960, 49
Arduino ADE, 6
Arduino Development Environment, 1
Arduino Quickstart, 3
Arduino team, 1
Artificial Neuron Network (ANN), 177
Attributes, 104
Axon, 150

B
Backpropagation, 183
Barometer, 46
Biological neuron, 150
BLE UUID, 36
Bluetooth Low Energy (BLE), 35

C
Color sensor, 52
Convergence, 166

D
Darlington configuration, 76
DC motor, 73, 75

DC motor ratings, 75
Decision trees, 104
Direct Sequence Spread Spectrum (FHSS),
    35
Duty cycle, 22

E
EFLL Library, 124, 130
Entropy, 106
Epoch, 178

F
Flash EEPROM, 21
Frequency hopping, 35
Fuzzy_controlled_robot, 139
Fuzzy logic, 123
Fuzzy_process, 124
Fuzzy rule development, 129
Fuzzy trapezoids, 126

G
Gesture detection, 49
Greenhouse, 36

I
Inertial Measurement Unit (IMU), 44
Input membership functions, 126
Inter–Integrated Circuit (I2C), 31
IR sensors, 79
ISM frequency band, 35

K
K Nearest Neighbors (KNN), 99

KNN color classifier, 99

L
LCD, serial, 24
LED biasing, 72
Light Emitting Diode (LED), 72
Linear actuator, 74
Linguistic variable, 129
Liquid Crystal Display (LCD), 24, 73

M
Maze navigating robot, 78
Mean square error, 178
Microphone, digital, 55
Motor operating parameters, 77
Multiple perceptron model, 167
Myelin, 150

N
Nano 33 BLE Sense, 17, 64
NINA B306 module, 20
NRF52840 processor, 17, 64

O
Output device, 72
Output_membership_functions, 129

P
Perceptron, 149
Perceptron challenegs, 176
Perceptron model, 150
Perceptron training, 151
Proximity sensor, 54
Pulse Width Modulation (PWM), 22, 77

R
Random Access Memory (RAM), 21
Relative humidity, 47
Robot IR sensors, 78
Robot platform, 79
Robot steering, 79
Root node, 105
Rosenblatt, 149
Run mode, 162

S
Serial communications, 23
Serial Peripheral Interface (SPI), 27
Servo motor, 74
Simulator, 58
Sketch, 7
Sketchbook, 7
Soma, 150
Stepper motor, 74
Strip LED, 11
Switch debouncing, 72
Switches, 70
Switch interface, 71

T
Timeline, 97
Tree traversal, 118

U
Unidirectional DC motor control , 76
Universal Synchronous and Asynchronous
        Serial Receiver and Transmitter
        (USART), 23

V
Volatile, 21
Voltage regulator, 68

Printed in the United States
by Baker & Taylor Publisher Services